# IS THIS WI-FI ORGANIC?

## A Guide to Spotting Misleading Science Online

### DAVE FARINA

mango

CORAL GABLES

For permission requests, please contact the publisher at:
Mango Publishing Group
2850 S Douglas Road, 2nd Floor
Coral Gables, FL 33134 USA
info@mangopublishinggroup.com

For special orders, quantity sales, course adoptions and corporate sales,
please email the publisher at sales@mango.bz. For trade and wholesale
sales, please contact Ingram Publisher Services at customer.service@
ingramcontent.com or +1.800.509.4887.

Is This Wi-Fi Organic? A Guide to Spotting Misleading Science Online

Library of Congress Cataloging-in-Publication number: 2020946575
ISBN: (print) 978-1-64250-415-6, (ebook) 978-1-64250-416-3
BISAC category code SCI080000, SCIENCE / Essays

Printed in the United States of America

*For my father, Vittorio, and my son, Reza.*
*May the thirst for knowledge perpetuate.*

# Table of Contents

## INTRODUCTION

# How Does the Public Perceive Science?

Let's play a game of word association.

# Science

Let this word simmer in your mind for a moment. Examine every texture. Taste the nuance. What does this word make you think of? How does it make you feel?

Do you imagine futuristic cityscapes? Do you feel hopeful?

Do you picture billowing smokestacks? Do you feel terrified?

Does it remind you of school? Does that hold a positive or negative connotation for you?

As to what you envisioned, it may have been any of the above, or something totally different still. But the general consensus of the American public on this matter can be quite easily traced through recent history. In the years following World War II, the American economy was booming. The middle class was gainfully employed, adequately fed and sheltered, and filled with optimism. Because of this, science was viewed as a gleaming obelisk of limitless advancement. It was Disneyland's Tomorrowland. It was the Jetsons. It was longevity and prosperity.

But over the decades that followed, this attitude changed. Administrations came and went. Post-war prosperity slowly dwindled, giving way to capitalism as we know it today, a poorly regulated industrial playground, absorbing wealth from the citizenry like a sponge. On this front, a resurgence of economic inequality suppressed what had briefly been a growing middle class. On another front, evidence of sustained environmental damage began to surface. Once these technology-enabled disasters came to outweigh the sheer excitement of the moon landing, that shiny obelisk was replaced with a barrel of toxic green goo. This trend in corporate practice is not exclusive to American corporations, nor is the resulting anti-establishment sentiment exclusive to the American public. These issues span the globe, infiltrating every nook and cranny of our civilization.

It is not surprising that there is widespread distrust of large corporations today. They are to blame for so much pollution and injury, and betrayal runs deep. But far too often, this distrust of industry is misplaced onto science itself. It is not just Big Business that is greeted with skepticism; the foundation of knowledge that commerce has built itself upon is also in question. We must make every effort possible to keep these two distinct entities separate. In one realm sits the undeniably valid scientific principles upon which technological progress rests, an ostensibly neutral body of knowledge. A separate realm contains all of the people and institutions that produce said progress, and the accompanying large-scale social change, for better or for worse. The latter camp must be judged for its actions on a case-by-case basis, without sullying the reliability of scientific knowledge, because while people have motives, information does not. In other words, although science denial is sometimes rooted in justifiable sentiment, it is not rooted in logic, and therefore simply is not the answer to our problems.

The dangers of rejecting our fundamental knowledge of nature are compounded in the information age, whose crowning achievement is the almighty internet. Although this invention has only been in widespread use for a mere two decades, it is already utterly and inextricably embedded in the functionality of our civilization. The internet is how we do things, and it is how we know things. It is how we communicate with one another. It is how we know which movies are playing this weekend, where to find the best Italian restaurant, or what the weather is like before looking out the window. Beyond trivialities, however, the internet is also the way many people probe the nature of reality to amass a worldview. Continuing the trend set into motion by the printing press nearly six centuries ago, the internet represents the ultimate democratization of information. There we find all of the information, updated in real time, essentially for free, apart from costly primary scientific literature. Anyone can access this mountain of data, and anyone can throw whatever they please onto the pile. While there are tremendous benefits to this paradigm shift, as it has become increasingly difficult to censor information, the transparency comes at a price. We have condemned ourselves to perpetually sifting through a digital cacophony of contradiction that often leaves truth obscured.

Prior to the internet, there were sources of information that were unanimously agreed upon to be trustworthy and reliable. Stories published by newspapers had to be heavily researched by professional journalists. Knowledge from an encyclopedia was not questioned by those who needed to reference a fact, because they were written by top specialists in every discipline, which contributed to their considerable cost. Whether we regard them as good or bad, those times are gone, and they are never coming back. Unlike the encyclopedias of old, the quality of information on the internet is not reliable. It ranges from outstanding to abysmal. For this reason, the internet can serve as a magic mirror, a place where people go to confirm pre-existing bias. Outlets that reflect what we already "know" are correct and trustworthy. Those that do not are ignored, deemed fraudulent, deceitful, paid for by malevolent institutions, or worse. This method of assessment rarely has any respect for the qualifications of those who produce the content we encounter, which has led to what is popularly referred to as the "post-truth era."

In this relatively new era, irrefutable facts and the firm consensus of the scientific community are often eschewed in favor of charlatans who peddle nothing more than a flashy narrative. Such pseudoscientific narratives have become popular for a number of reasons. On a more subtle and philosophical level, we are susceptible to wishful thinking. Science offers us a cold and indifferent universe, a narrative that is antithetical to the divinely ordained status humanity has grown accustomed to over the millennia. But the appeal to pseudoscience also arrives largely in response to harmful practices enacted by the aforementioned large corporations. And because anti-corporate sentiment is so widely held, it has become trivial to fool large sectors of the population with a handful of buzzwords and scare tactics.

There was a time when charlatans would travel from town to town selling snake oil. This useless concoction was peddled as a cure-all for what ails you, and gullible people would buy it. Since that time, snake oil has become a euphemism for deceptive marketing, and the sale of digital snake oil is rampant on the internet. This new and improved incarnation is not limited to potions and lotions. It has morphed into an idea, a feeling, a way of believing the universe must operate.

It has always been clear that if a number of people are willing to buy a particular product, someone will create that product in order to profit off of the demand. But beyond this, it is now clear that if enough people simply want to believe in a particular reality, media will be crafted to corroborate that reality. The desire to believe becomes the tendency to click, and clicks mean bucks. This is no secret. Information exists on the internet because it is to be clicked upon, and not because it necessarily correlates with reality. Therefore, finding what you want to believe on the internet doesn't immediately make it true, no matter how true it feels. This aspect of the post-truth era is a huge problem. Because almost every opinion imaginable exists on the internet, whether it is based in fact or not, even highly intelligent users can regularly find themselves incapacitated.

This product that I'm interested in purchasing, is it safe to use? One source says no, another says yes. Others, something in between.

Is this corporation behaving ethically? One source says no, another says yes. And many more are somewhere in the middle.

What do I want? What should I do? Who should I be? Again, a bevy of bumf.

This peril is compounded by the fact that misinformation is shared on a daily basis, such that it spreads like a virus across countless social media platforms, rapidly gaining momentum in a way that was impossible prior to the existence of the internet. So where does that leave us? If every worldview imaginable is supported in one place or another, how can we sort fact from falsehood? Is there some reliable algorithm for sniffing out all the lies and propaganda, or are we doomed to exist in epistemological limbo for the remainder of human civilization?

Well, there is a way, but it involves learning. To some, the prospect of learning something new is invigorating. To most, who are exhausted after a long day of work, learning something new is a chore and a nuisance. But there really is no other way. If you want to know whether some product is safe to use, it will not suffice to simply read yes or no on the internet, as they are both there for you to find. You must be equipped to comprehend the explanations that are given. Some of them will correlate with the nature of reality, and others will not. In many cases, this is shockingly trivial to discern.

To be scientifically literate, you don't have to be an expert in anything. You don't have to know all of the science. In fact, no one does, not by a longshot. The era of polymaths like Benjamin Franklin and Leonardo Da Vinci is long gone. There is far too much science for anyone, even scientists, to know any more than a tiny sliver of the pie. It would seem that the Renaissance men of Da Vinci's caliber were truly only possible in the Renaissance, prior to the landslide of progress that was the twentieth century. Today we are left with a behemoth for a body of scientific knowledge. But the immensity of this beast should not discourage anyone from examining it. It is far too common that individuals label themselves as "non-science people," concluding that it's all just above their intellect. But the average citizen doesn't need to understand quantum mechanics. We don't even have to do any math. We simply need a conversational level of proficiency in the most rudimentary concepts from the major areas of scientific inquiry.

In this book there are tools. Basic information will be offered regarding a few fundamental fields, those being chemistry, biochemistry, biology, and physics. This knowledge will then be used to deconstruct popular narratives found in various media. We will look at claims that are made out there in the ether and evaluate them. Do they hold up to scrutiny? If not, then precisely how much do we need to know in order to recognize such false information? Again, you may be surprised at just how little knowledge is required in order to read something and instantly reject it, as those who peddle pseudoscience underestimate you. They arrogantly believe that they can fabricate narratives which are completely inconsistent with what we, as a species, know to be true about the universe, and the vast majority of readers will never be the wiser.

Let's prove them wrong.

CHAPTER I

# What Are All These Lines and Hexagons?

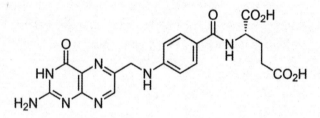

Folic acid, displayed in line notation.

Take a look at the image above. Don't panic! That's a **molecule**. As it happens, this molecule is called folic acid, also known as vitamin B9, which is an essential component of most living things. It may just look like some lines, shapes, and letters, but that's the way this particular molecule is visually represented. Molecules are collections of **atoms**, and atoms make up pretty much everything you see and touch on earth. If we want to be able to talk about what goes on inside the human body, what drugs are and what they do, and how all of this intersects with industry, this is where we have to start. If we wish to separate myth from reality, we have to be fundamentally equipped to understand why organic does not mean pure, natural does not mean good, synthetic does not mean bad, and all of the other common misconceptions that make us vulnerable to false advertising. In order to later dig into all of this juicy stuff, which is the primary purpose of this book, we first have to be able to look at a structure like this and know what it represents. That means we are going to have to learn a few things about chemistry.

I can sense you recoiling already. What did you get yourself into? Is this a concise and leisurely airplane read, or is this homework? Fret not. We are simply going to breeze through the bare minimum background information as fast as humanly possible. Flip ahead a few pages if you don't believe me. In less time than it would take to cook a frozen pizza, you will be able to comprehend the above image, so let's get started.

An introduction to chemistry must begin with atoms. Atoms, though incredibly tiny, are made of three smaller things still. Those are protons and neutrons, which are of similar size and mass, sitting in the central nucleus, and electrons, which are even tinier, existing far away from the nucleus. Protons have a positive electric charge, electrons have a negative electric charge, while neutrons are neutral, bearing no electric charge.

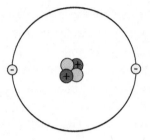

A crude representation of a helium atom, with protons in red,
neutrons in green, and electrons in yellow.

Oppositely charged particles attract one another, and like
charges repel. The protons and electrons, through this
influence upon one another via the **electromagnetic force**, are
responsible for the entirety of chemistry. Meanwhile, neutrons
do not serve this function. Instead, they help hold the atomic
nucleus together via a different force called the **strong nuclear
force**. This force is much stronger than the electromagnetic
repulsion pushing the protons apart, and is thus the reason
atomic nuclei are stable, with the protons and neutrons fused
together like a bunch of grapes. So that's where we begin,
with protons, neutrons, electrons, and the atoms they form by
their various combinations. We could talk about even smaller
particles still, or we could talk about precisely why opposite
charges attract, but it's all too complicated and doesn't serve
our main purpose, so just take these facts for granted and we
will build from here.

Now, different atoms can contain different numbers of these
particles. Take the proton. An atom might have one proton in

its nucleus. It might have two, or three, all the way up to over a hundred. It will also have some neutrons, which for smaller atoms will be the same or almost the same as the number of protons, and for larger atoms, closer to one and a half times the number of protons. But it is specifically the number of protons that determines which element the atom belongs to, which essentially means the type of atom, and every element has a corresponding set of chemical properties. An atom with one proton is called hydrogen, represented by the symbol **H**. An atom with two protons is called helium, represented by the symbol **He**. Each time we add a proton, we get the next element on the ever-familiar periodic table, each with its own characteristic symbol of either one or two letters, displayed below an **atomic number** that refers to the number of protons in the nucleus.

The periodic table of the elements.

Adding to this, a neutral atom has precisely as many electrons as it does protons. These electrons move at astonishing speeds within little regions of space called orbitals that project away from the nucleus. If we think of an atom as an apartment complex, then an orbital is like one of the apartments where electrons can stay, up to two of them per unit. These orbitals have different shapes which require knowledge of quantum mechanics to understand, so we won't get into that here. We simply must know that electrons are the particles that allow chemistry to occur.

> **TL;DR—A chemical element is a particular type of atom, defined by the number of protons in its nucleus.**

$$CH_4 \; + \; 2O_2 \; \longrightarrow \; CO_2 \; + \; 2H_2O$$

The combustion of methane.

Next we must understand that atoms can bond with one another, like two fingers in a Chinese finger trap, and when these bonds break and form, that's chemistry in action.

**Chemical reactions** are processes in which atoms rearrange their combinations with one another. Visualizing atoms now as spheres of different colors, refer to the example shown, where methane ($CH_4$) reacts in the presence of oxygen ($O_2$) to produce carbon dioxide ($CO_2$) and water ($H_2O$). Each of these substances is a molecule, which contains atoms of one or more elements participating in chemical bonds with one another. When we write these, each number in subscript refers to the number of atoms present of the element listed immediately to its left, so methane contains one carbon atom and four hydrogen atoms. Coefficients to the left of a substance indicate the number of molecules of that substance involved in the reaction, such as the two water molecules produced. In the course of any chemical reaction, atoms rearrange their configurations to produce new substances, and in doing so, they conserve their number, so no atoms are created or destroyed.

But what is a chemical bond? Well as previously mentioned, electrons and protons come together to form atoms because of the electromagnetic attraction between them, due to their opposing charges. Beyond this, protons from the nucleus of one atom can also attract electrons from another atom. If two atoms each share one electron with the other, such that the two shared electrons sit between them, feeling the pull of each nucleus, this constitutes a **covalent bond**, which we typically represent in shorthand as simply a line between two atoms.

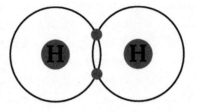

A hydrogen molecule ($H_2$) contains one covalent bond, which involves two shared electrons, shown here sitting between the two hydrogen nuclei.

Two atoms can also share two pairs of electrons to make two bonds, which we would represent with two lines, and refer to as a double bond. They can even share three pairs of electrons to make three bonds, which we call a triple bond. The greater the number of shared electrons, the stronger the bond.

Representing single bonds ($C_2H_6$), double bonds ($C_2H_4$), and triple bonds ($C_2H_2$), with covalent bonds illustrated simply as lines between atoms.

And finally, atoms can possess pairs of electrons that do not form a bond, but rather belong exclusively to one atom, which we would call a lone pair of electrons.

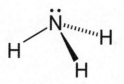

Ammonia (NH$_3$) contains three single covalent bonds, and one lone pair of
electrons on the nitrogen atom. The dash and wedge bond notation will be
explained shortly.

*TL;DR—Atoms share electrons to form covalent bonds.*

Once again, these types of chemical bonds we are discussing
are called covalent bonds, which refers to bonds in which
electrons are shared between two atoms. There are other types
of bonds as well, which will be discussed a bit later. But for now,
certain elements that are primarily found in the upper-right
section of the periodic table, which are called non-metals, are
very good at making covalent bonds with one another. These
include hydrogen, carbon, nitrogen, oxygen, phosphorus, and
sulfur, which also happen to be the six elements most commonly
found in living organisms. Atoms of these elements come
together in different combinations to form countless different
molecules, both familiar and unfamiliar, so let's now get a better
understanding of the properties of molecules.

Linear molecular geometry.

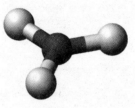

Trigonal planar molecular geometry.

Tetrahedral molecular geometry.

Looking first at small molecules, like the ones in the combustion reaction we mentioned earlier, these take on specific geometries. Consider some central atom with multiple atoms or sets of atoms attached to it, which we can call groups. These groups will spread out in space so as to be as far away from each other as possible, because the negatively charged electrons surrounding those atoms repel one another. Recall that while particles of opposite charge experience an attraction, particles of like charge experience a repulsion, so imagine like poles of tiny bar magnets pushing themselves apart. If there are two groups surrounding a central atom, we get a line, as they can get no farther away from each other than being on opposite sides of the central atom. If there are three, we get a triangle, which again maximizes these bond angles. If there are four, we begin to utilize the third spatial dimension, and we get a tetrahedral arrangement. It is the repulsion between these groups that is pushing them as far away from each other as geometry will physically allow. To be clear, a "group" in this context can be either an atom or a lone pair of electrons, as these will produce repulsion in a similar manner. This is the reason that molecules like carbon dioxide and water have different shapes, despite each containing three atoms.

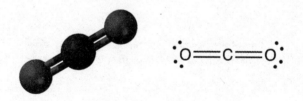

Carbon dioxide ($CO_2$) exhibits linear geometry.

Water ($H_2O$) exhibits tetrahedral geometry.

With carbon dioxide, the carbon atom is surrounded by two groups, those being the two oxygen atoms, and will thus be a linear molecule. Water involves an oxygen atom surrounded by two hydrogen atoms and two lone pairs of electrons, which equals four groups. This results in a tetrahedral arrangement and a bent shape for the molecule.

> *TL;DR—Molecules take on specific shapes that allow groups to spread out as much as geometrically possible.*

Now let's allow these molecules to get a little bigger. Molecules will commonly contain primarily carbon and hydrogen, so let's go ahead and construct a molecule with just those elements, which we would call a hydrocarbon. Hydrogen has only one electron, so it can make only one bond, in this case by sharing it with a carbon atom. Carbon has four so-called **valence electrons**, which means the outermost ones that are available for bonding, so carbon can make up to four bonds. Here we can see that each carbon makes four bonds to four different atoms. This includes either one or two other carbon atoms, with the remaining bonds going to hydrogen atoms. Since each carbon makes four bonds, they exhibit the tetrahedral geometry we previously mentioned, and that's why we see this zig-zag pattern. It's also why we have to use what are called dash bonds and wedge bonds, so as to display the molecular geometry accurately. The dashes mean that the bond projects away from the viewer, beyond the page, while the wedges mean that the bond projects toward the viewer, in front of the page. This makes it possible to represent three-dimensional molecules on a two-dimensional page or screen. And with that understood, we can observe the true geometry of a molecule like this.

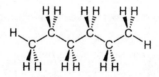

Hexane ($C_6H_{14}$), a hydrocarbon.

Now that we have a basic understanding of chemical bonding and molecular geometry, we can translate this structural representation into what we call line notation, which is the method by which that molecule at the beginning of the chapter was depicted. To do this, we need to know just two things. First, when writing molecules in line notation, every vertex and endpoint represents a carbon atom, since most large molecules are built from a carbon skeleton. And second, the hydrogen atoms in the molecule are implied, meaning they are not explicitly drawn. We simply have to know that they are there. So, if we translate hexane into line notation, we are simply left with this zig-zag line.

Hexane in line notation.

Once again, every point is a carbon atom, and however many bonds are missing from each carbon, that's the number of hydrogen atoms that are assumed to be there. So the carbon on one end shows just one bond to the carbon atom next door, and that means three bonds to three separate hydrogen atoms are also there but not shown, since carbon tends to make four bonds. The carbons in the middle of the chain each show two bonds, one to each of the carbons on either side, and that means they each have two bonds to two separate hydrogen atoms that are not shown, again to get to a total of four bonds.

*TL;DR—With line notation, every vertex and endpoint is a carbon atom, and the hydrogen atoms are implied.*

Cyclohexane in line notation.

Molecules aren't always linear. Sometimes they can be cyclic, with the carbon chain looping back around on itself. While this looks like a simple hexagon, we now understand that this is actually six carbon atoms, each of which is bonded to two hydrogen atoms. This symbol represents a molecule called cyclohexane, where the prefix "cyclo" indicates a cyclic molecule, and the prefix "hex" indicates six carbon atoms, while the suffix "ane" indicates that there is only one bond in between each of the carbon atoms. As we mentioned, atoms of certain elements, especially carbon, can make more than one bond between them. These are called double bonds and triple bonds, and in line notation we continue to represent these with multiple lines. So taking cyclohexane and changing three of these single carbon-carbon bonds to double bonds, we get another molecule called benzene, which acts as the basis of many other molecules. Bear in mind that instead of two hydrogen atoms being implied on each carbon, there is only one, as each carbon is showing three bonds instead of two, leaving only one other bond available for bonding to hydrogen.

Benzene in line notation.

With all of this understood, recall the image of folic acid at the beginning of the chapter. Suddenly things are a bit less cryptic. Every unlabeled vertex is a carbon atom, while atoms of other elements are represented by their chemical symbols. In this case that means nitrogen, oxygen, and any hydrogen atoms which are bound to elements other than carbon. Every line is a bond, and two lines between two atoms means a double bond. Sometimes to save space we write common groups all at once, so OH means oxygen bonded to hydrogen, and we simply omit the line. The same goes for NH, where the line from N to H happens to be omitted as a shorthand. And most of the molecule is pretty flat, except for the one dash bond we see, which pushes that part of the molecule beyond the plane of the page.

Now that wasn't so bad, was it? Granted, we skipped a lot of information, and for the information we did cover, we went with the ultra-condensed version. But at the very least, we can now encounter molecules in line notation and know what we are looking at, even if we aren't yet familiar with what molecules do. Now let's take a brief moment to talk about that as well.

As was stated earlier, molecules can undergo chemical reactions, whereby atoms rearrange their combinations. This will happen when two molecules collide with enough speed so as to react, and in a particular physical orientation. So what is this orientation? Well let's talk some more about electrons. Recall that the lines between the atoms are covalent bonds, whereby the two atoms are sharing two electrons. But these are not always shared evenly. One atom can be said to be more **electronegative** than the other, which means that it holds electrons more tightly, and thus the electron density in that bond will skew more closely toward the more electronegative atom, like they are playing tug-of-war with an electron rope and one atom is winning. This leaves one atom with some excess electron density, and thus a partial negative charge, while the other atom is left with some electron deficiency, and thus a partial positive charge. We call this a polar covalent bond, where **polar** refers to the polarization of the electron density, which rests closer to one atom than the other to some degree. When collisions occur which allow opposing partial charges on separate molecules to make contact, this is an example of a situation that is more likely to result in a chemical reaction.

Once again, we know that opposite charges attract, and this is truly the driving force behind every chemical phenomenon. Chemical reactions occur when some electron excess meets some electron deficiency. When minusness meets plusness, so to speak. If a collision occurs such that these two regions can interact, new bonds replace old bonds, atoms rearrange, and we end up with different molecules than what we started with, provided that this new configuration is preferential to the original, according to a parameter called Gibbs free energy. The energy required to break the bonds is supplied by the collision itself, and then energy is released when the new bonds form. Chemistry has occurred.

> *TL;DR—Chemical reactions occur when some electron excess, or minusness, meets some electron deficiency, or plusness.*

As is abundantly clear at this point, there is only one force that matters in chemistry, the electromagnetic force. It is the reason atoms exist, due to the attraction exhibited by subatomic particles of opposing charge. It is the reason those atoms come together through chemical bonds to form molecules. It is the reason molecules take on particular shapes, due to the repulsion of electrons with like charges. It is the reason electron density distributes itself around a molecule in a particular way. And it is the reason these atoms rearrange themselves into new combinations when molecules collide with sufficient energy and in the correct orientation, a phenomenon which we call a chemical reaction. Electron-rich meets electron-poor, a tale as old as time. And with that tale told, we know just enough about chemistry to move forward into other territory.

# CHAPTER 2

# **The Death of Vitalism**

Imagine a time prior to the nineteenth century, a time when modern chemistry does not exist. Mankind has very little comprehension of the composition of matter. Atoms are merely an unsubstantiated philosophical concept, a relic from ancient Greek thinkers. Alchemy, the pseudoscientific practice of attempting to transform one element into another, is alive and well. In short, in this particular time, we have no clue what things are fundamentally made of. Because of this, a seemingly plausible belief called "vitalism" is widely held. This is the notion that the things which make up living organisms, all the bodily flesh and fluids, are somehow special and different from other matter. They possess some kind of intrinsic vitality that is absent in inanimate matter and beyond the capacity of mortal man to produce. These substances are bestowed with this vitality by God, and as man is not God, man is hopelessly unable to produce any substance which is of life.

However, this notion was dispelled in the early nineteenth century, through the work of a number of chemists of the era. One major achievement toward this end took place in 1828, when the German chemist Friedrich Wöhler accidentally synthesized the compound urea. This is a small molecule produced by mammals in order to excrete unused nitrogen atoms, which is done through urine. Chemists of the time were able to make this molecule in the laboratory from other "non-vital" substances, such as ammonium cyanate. What was especially surprising was that urea made in the lab was completely identical to "natural" urea in absolutely every respect. This result, which was not immediately comprehended at the time, demonstrates that there is no special property to the stuff of life that is beyond our mortal reach. The chemistry of life is the same as the chemistry of non-life. And with that, the field of organic chemistry was born.

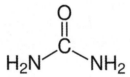

Urea, synthesized by Friedrich Wöhler in 1828.

*TL;DR—There is no special "vitality" held by the molecules in living organisms that makes them fundamentally different from other molecules.*

Over the decades that followed, it became increasingly clear through the work of other chemists that there was no fundamental barrier to the synthesis of any particular molecule, and as their skills in the laboratory became more sophisticated, chemists developed increasingly clever strategies to synthesize molecules that were progressively more structurally complex, a practice which persists to this day. Our current prowess in the synthesis of molecules is astonishing. As an example, observe the structure of palytoxin, a compound found in certain types of coral, whose synthesis was achieved by Professor Yoshito Kishi at Harvard University in 1994.

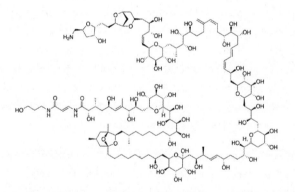

Palytoxin, synthesized by Professor Yoshito Kishi in 1994.

In fact, all of the major molecules of life, including sugars, proteins, nucleic acids, and vitamins, all of which will be discussed later, can easily be synthesized in the lab using strategies invented by chemists. When these molecules are inserted into a living system, they function precisely as those made by nature. There is absolutely no difference. However, as conclusive as this fact may be to the scientific community, it continues to elude large sectors of the public even two centuries after Wöhler initially uncovered it. Many people do not seem to fully understand or appreciate the fact that chemists can make molecules that exist in nature, and that they are absolutely identical in every way conceivable, because the properties of a molecule are completely defined by its structure, and not at all by its origin. There is no "vital force" or "vital energy" that exists so as to be present in one version and not the other. The misunderstanding of this fundamental truth leads to a plethora of problems, which will be a persistent focus of chapters to come. But before we get there, we need to continue the thread we've just begun. It will not suffice for me to simply tell you that two molecules are the same. We have to understand precisely what chemists do, and how they do it, in order for this conclusion to be blatantly obvious and self-evident. We need to know the mind of an organic chemist.

What does an organic chemist do? What does **organic** mean, in the first place? With respect to the masses, this is a buzzword that begs to be deconstructed for hours on end, and we will examine the word in that context a bit later. But in science, the word organic has a specific and straightforward meaning. An organic molecule is one that is carbon-based. Carbon has to be in there, whether it's one atom or a thousand. Hydrogen is almost always in there too, while oxygen and nitrogen are the next most common participants, followed by a small handful of other elements. To qualify as organic, a molecule need not be found in a living organism. It need not be made by natural processes. It doesn't even have to exist at all, it could be something conjured up in the mind of a chemist, never before encountered in the entire universe. If it's carbon-based, it's organic, and that's all there is to it. A synthetic organic chemist, which actually is not an oxymoron, is a person who synthesizes organic compounds.

> *TL;DR—In chemistry, "organic" just means carbon-based, and nothing else.*

So how does one do this? How does one make molecules? Does this involve the smallest tweezers imaginable and the delicate touch of a surgeon? Or one of those people that builds tiny ships in bottles? Nothing of the sort, actually. We can't see molecules. They are just too small. To give you an idea of how small they are, go and get a glass of water. Now stare at it. Imagine all those water molecules sloshing around, bumping into one another as they go. How many are in there? A million? A billion? Actually, it's closer to a trillion trillion.

What does that even mean? Well, a trillion is a million millions.

So imagine a million of something. Perhaps, oh…why not apples? Picture them all together, arranged in a cube that is a hundred apples long, a hundred apples wide, and a hundred apples tall, forming some bizarre avant-garde art piece at an elitist state fair.

Now imagine not just one cube of a million apples, but a million such cubes, each with a million apples. That's a trillion apples. Got it? Now you're imagining a trillion.

Now combine all one trillion of those apples into one enormous pile. Then imagine a trillion of those piles.

Does your head hurt yet? That's because this number, the number of molecules in your glass of water, is larger than the number of stars in the observable universe. By a lot. The number of molecules that comprise any macroscopic object defies human comprehension. It's too large for us to fathom.

> *TL;DR—Molecules are mind-bogglingly*
> *small and numerous.*

So in short, we do not build molecules the way we build IKEA furniture. We don't build them one at a time. Instead, chemists will coax astronomical numbers of molecules into doing what they want them to do, all at once. This is achieved by putting molecules in some kind of inert glass vessel and manipulating their environment. They are dissolved in a suitable solvent. The surrounding temperature can be made very hot or very cold. These molecules are then carefully introduced to new molecules, with which they may react. Whatever we do, we are taking some particular molecule by the quintillions and inducing a specific transformation on all of them at once, by understanding the principles that dictate what molecules do. We can transform one part of a molecule into something else. We can cause two molecules to connect in a specific place and form a larger molecule. Obviously this process has become extremely sophisticated over the decades, in a way that would take several years in the classroom to fully appreciate, but on the most fundamental level, that's it. We put some molecules in a flask and do things to them to turn them into other molecules. Furthermore, we are able to do this with an efficiency and ingenuity that eludes nature, due to one key factor. We have sentience. Nature makes no decisions, it is simply a set of rules, and molecules will react as they must, according to localized conditions and what is in their immediate vicinity. Humans, however, are not relegated to blind chance, nor limited to nature's typical building blocks. Chemists can deliberately elect to combine any set of compounds they wish, and under any conditions. We do not have to confine ourselves to the narrow temperature and pH range inside the body, or in the ocean. We can use extreme temperatures and highly reactive compounds which would destroy any biological system. We are not limited by the relative abundance of the elements. We can mine for rare earth elements, make interesting things with them, and toss them in the pot. Through the agency of a sentient chemist, nature does chemistry that is more subtle and directed than it has ever done before.

With all of this, the crucial thing to remember is that nature also builds molecules. There is a tendency amongst the public to envision chemical synthesis as the work of a mad scientist, cobbling together a Frankenstein's monster which lacks the humanity of the entity that inspired it. This could not be further from the truth, and as was promised, an investigation surrounding the details of this narrative will be the focus of the next chapter. But before we get there, let's tackle one other misconception.

With some understanding of what molecules are, we can begin to discuss why one might be bad or good for you. We will dig a lot deeper into this concept after we learn more about the structures found inside the body, but a basic principle to be aware of is that molecules do what they do strictly because of their composition and shape. There are certain groups of atoms called **functional groups** that are commonly found in molecules, which behave in particular ways and contribute to the reactivity of a molecule. In addition, the overall three-dimensional shape of a molecule determines what kinds of nooks and crannies it fits into within the body, in the way of interactions with other, larger molecules. These functional groups consist of atoms of the common elements, but their specific arrangement is important. What an oxygen atom does when bonded to a hydrogen atom is different than what it does when bonded to two carbon atoms, or another oxygen atom, and so forth. So there is a big difference between an element and an atom of a particular element that is found within a molecule.

*TL;DR—A molecule that contains one or more atoms of a particular element will not behave the same way as that element in its elemental form.*

Consider mercury, for example. Elemental mercury, which is the stuff inside thermometers, is bad for you. It has a rather high toxicity. Let's quickly define this concept. **Toxicity** describes how harmful a substance is to our health, which is typically quantified by a parameter called $LD_{50}$, where LD stands for lethal dose. This is the amount that will kill fifty percent of some test population. Literally any substance can be assigned such a value, because any substance will be harmful in high enough dosage, even water. Believe it or not, it is possible to die from drinking too much water, and I'm not talking about drowning. Even though we need to drink water to live, drinking too much water causes water intoxication, which disrupts brain function, and can be fatal. Of course, you have to drink a heroic amount of water for this to happen, so don't start getting paranoid when you go to fill up your water bottle. But the same thing applies to any other compound. Without oxygen, we die in minutes. And yet, breathing in pure oxygen for a few hours will kill you. There is some degree of consumption or exposure at which point harm will be done, no matter what substance we are looking at. On the other end of the spectrum, there exist classes of molecules whereby even a microscopic quantity in the bloodstream can be lethal, for one reason or another. To speak about toxicity as though it is a property that some substances have and others do not is simply inaccurate.

> **TL;DR—Literally everything is toxic in high enough dosage.**

Therefore, as sensible as it may seem at first glance, we must get over the notion of "bad chemicals" and "good chemicals." Anything can kill you if the dosage is high enough, but some things are worse than others, due to factors that are specific to the biology of the particular organism in question. Once again, mercury is pretty bad for us in its elemental form, but this does not apply equivalently to any compound that contains a mercury atom. Take for example a compound called thimerosal. As can be seen from its structure, it contains one mercury atom, which is abbreviated as **Hg**. This compound has had application in vaccines as a preservative to prevent contamination, and it is not harmful to the human body at the exceptionally low dosage that is utilized in vaccines. But anti-vaccine activists refer to this mercury-containing compound as though it were the same as elemental mercury in terms of toxicity, which is objectively false. The way a singular mercury atom interacts with the body is in no way related to the way a molecule with a mercury atom in it will interact with the body, for biochemical reasons that will become increasingly clear in upcoming chapters.

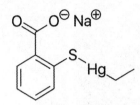

Thimerosal contains one mercury atom.

This notion is quite easy to elucidate by looking at a much
simpler example. Sodium is a highly combustible metal that
ignites spontaneously when exposed to moisture and causes
severe irritation upon contact with the skin. Chlorine gas is
highly toxic, and exposure can be fatal in even moderate
amounts, which is why it was used as a chemical weapon in
World War I. Two elements, both considerably harmful. What
do you suppose happens when we put them together? When
we combine sodium and chlorine, we get completely harmless
sodium chloride, otherwise known as table salt. No combustion,
no irritation, and zero utility as a chemical weapon. In fact,
this substance is absolutely essential to the human body for a
number of cellular processes. And beyond this, who could go a
week without a salty snack? As anyone can see, sodium chloride
is not the same as elemental sodium and elemental chlorine.
The reason for this dramatic shift in properties is due to the way
these elements interact to form this particular compound, which
is distinct from other compounds.

Sodium metal and chlorine gas are both considerably harmful.

Sodium chloride, also known as table salt.

To briefly summarize the chemistry concept that will elucidate this previous example, we must understand that chlorine atoms can gain one electron to fill up their outermost shell and become more stable. Chlorine atoms that gain an electron are called chloride ions, where an **ion** is any atom with an overall non-neutral charge. Sodium atoms, on the other hand, have only one electron in their outermost shell, which makes it easy to lose. So when they encounter one another, sodium

will readily transfer its lone valence electron over to chlorine
to become a sodium ion, and everyone wins. We are left with
a positively charged sodium ion, since it now has one more
proton in the nucleus than it does electrons, and a negatively
charged chloride ion, since it now has one more electron than it
does protons in the nucleus. And as we said, plus and minus like
to interact, so we end up with a grid of these ions in a repeating
lattice structure, which we call an **ionic solid**.

Sodium chloride is an ionic solid.

This is a bit different from the covalent compounds we have
been discussing, but it illustrates this important principle all
the same. It does not suffice to simply look at which elements
are present in a compound to assess its toxicity, or any other
property. We must know how these elements are arranged.
There are plenty of extremely toxic compounds that are
comprised of merely carbon, hydrogen, oxygen, and nitrogen,
which are precisely the same elements that are present within
all the other compounds that are found in the body. It is their
three-dimensional arrangement that bestows them with a

particular bioactivity, which we will come to understand more thoroughly soon enough. But before we dive into the human body, we have more to discuss regarding molecules, so let's tackle a few more misconceptions.

CHAPTER 3

# Natural vs. Synthetic (Tackling Chemophobia)

Now that we have a basic idea of what chemists do, we are ready to tackle an enormous misconception that is widespread amongst the public. This is the notion introduced in the previous chapter, that there is some inherent difference between a molecule made by natural processes and a molecule made in a laboratory. And not only this, but also the belief that anything natural is inherently good, and anything synthetic is inherently bad. This way of thinking is so spectacularly and fundamentally false, that it is difficult to know where to begin in unpacking the narratives and influences that have produced this sentiment. But perhaps we can start by thoroughly dismantling this falsehood from top to bottom.

Let's first recall the work of Wöhler and other chemists in the early nineteenth century. Even though it was demonstrated in this time that humans can make the molecules that nature makes, an echo of vitalism remains alive and well today. But as we have come to understand, this worldview neglects to consider the fact that nature does chemistry too. Nature must assemble molecules through chemical reactions just like we do, and no matter which pathway is taken, the end result is the same. A molecule is just some specific atoms arranged in a specific way, and any occurrence of those atoms arranged in that way is that molecule, without exception. Atoms of a particular element can't behave differently from one another in general, let alone according to some specific origin or experience, as atoms don't have memories. Atoms don't know where they are from or where they have been.

Those who would disagree undoubtedly place false connotations onto these words, natural and synthetic, so let's make sure to define them. **Natural** simply means made by nature. It does not mean good, it does not mean pure, it does not mean anything other than that it came to be by natural processes, without sentient intervention. **Synthetic** simply means made by humans or human-built machines. It does not mean bad, it does not mean inferior, it does not mean anything other than that it came to be through deliberate human action. Any molecule could be synthetic. If someone were to mix samples of hydrogen gas and oxygen gas and then introduce a spark, they would produce water, and this water could be referred to as synthetic. It's still just two hydrogen atoms and an oxygen atom, but we forced it into existence.

$$2H_2 + O_2 \rightarrow 2H_2O$$

The reason that these definitions have no relevance beyond this singular context is that biological systems have no way of determining where a molecule has come from. We can synthesize pheromones that attract insects identically to their own copies. We can synthesize antibiotics that are naturally made by molds which kill bacteria in precisely the same way as theirs do. We can synthesize ethylene and use it to cause plants to bloom and fruit to ripen, mimicking the plant's own mechanisms. That nature can't tell the difference, thereby subjecting herself to manipulation, is a compelling demonstration of this principle.

> **TL;DR—Natural does not equal good, and synthetic does not equal bad.**

Therefore, by looking at natural and synthetic versions of the same molecule, whether it's water, a vitamin, or anything else, we can see how silly the mantra "natural, good; synthetic, bad" really is. The most toxic compounds known to man occur in nature. These are found in plants, as well as animals like frogs and snakes, essentially qualifying as biochemical weapons that can kill a person in a few seconds from minimal exposure. These weapons were stumbled upon by nature, not by humans.

Nature is ruthless. A single gram of the botulinum toxin from the bacterial species *Clostridium botulinum* is enough to kill ten million people by injection, making it the most powerful poison known to man. It is tempting for many people to view nature as an eternally sunny meadow, but nature is also death and destruction. Nature is pestilence, it is disease, it is natural disaster. Nature is both life and death, it is a dichotomy. In the internet age, a portrayal of nature as an eternally sunny meadow is merely a marketing tactic. It hits some fuzzy region of the brain that makes us feel safe and cared for, with no mention of the naturally occurring compounds that epitomize lethality. Conversely, synthetic molecules include drugs that save lives, as well as completely identical versions of all the naturally occurring molecules, whether bad or good. So we can safely reject this falsehood and understand instead that molecules must be examined on a case-by-case basis to determine their properties, regardless of their origin. In short, what a molecule does in the body, or its bioactivity, has nothing to do with how that molecule came to be.

The irony of such a false narrative is compounded when we look at another misused word, the ever-familiar toxin. So many products and services boast cleansing properties, promising to rid your body of toxins. This is a real word with a specific scientific meaning, and it is quite regularly used improperly. A **toxin** is, by definition, a poisonous substance produced within a living organism. So while we can say that many toxins are highly toxic, and some not quite as much, we must understand that they are always natural in origin. In popular media, however, this has become a buzzword that alludes to a harmful chemical produced by a malicious scientist in a lab coat. It conjures up imagery of greedy industries run amok. This connotation is exploited for profit, targeting those who don't understand that if you possess a liver and two kidneys, your body is doing all the detoxification it can ever or will ever do. No smoothies and no amount of sweating can change this, no matter how delicious or invigorating these products and practices may taste or feel. The fallacious notion of toxin-cleansing is just another symptom of the "natural, good; synthetic, bad" way of thinking.

> *TL;DR—Anyone that is trying to sell you methods of "detox" either has no clue what they are talking about, or is deliberately manipulating you, so save your money.*

But beyond rejecting this false narrative, we want to understand why it exists in the first place. If it isn't true, why do people think this way? It seems apparent, at least to me, that it is due to a simple linguistic coincidence. The word "synthetic" has many connotations, and there are certain substances, such as fabrics and other materials, for which synthetic versions truly are shoddy imitations of some natural counterpart. There is a material derived from something in nature, and there is a cheaper knockoff, not quite the same but similar, and arguably inferior. This realm is where the negative connotation for this word is born. It is associated with "artificial," and regarded about as desirable as an artificial hip. But we must understand that these examples, whether the natural fabric and the synthetic knockoff, or the biological hip and the artificial hip, are different materials, or even different objects, altogether. This connotation does not transfer to the molecular world. We now know what the word synthetic means for molecules, and we know that it does not have inherently negative implications. So when we see outlets that abuse this term, we ought to conclude that we are being manipulated.

There is a related term we have been using that must also be discussed, and that is the word "chemical" when used as a noun. Some commonly uttered phrases:

*Are there any chemicals in this?*

*I don't like to use products that have chemicals in them.*

This word sits in tandem with "synthetic" as eliciting doubt and fear. Distilled to its purest form, the narrative could be summarized as follows:

*Chemicals are synthetic, and synthetic is bad, so chemicals are bad.*

But **chemicals** are just molecules. The terms are essentially synonymous. Everything around you is chemicals. Water is a chemical. Oxygen is a chemical. This word, like the word synthetic, got a bad reputation somehow, and it must be rehabilitated. The most likely explanation is that a discussion of chemicals which laypeople commonly engage in regards the danger of specific harmful chemicals, and thus the connotation evolved by repetition. But the chemicals that clean your floor are not the only chemicals, everything else is chemicals too. The irrational fear of chemicals and chemical-sounding things is called **chemophobia**, and it is indeed a phobia like any other, which is derived from our fear of the unknown or unfamiliar. There are those who exploit chemophobia to sell us exorbitantly priced products, just the way that there are those who exploit xenophobia, which is a fear of people from other countries, for self-serving political purposes. It is manipulative, plain and simple, so don't succumb to molecular discrimination.

> *TL;DR—Pretty much everything is chemicals.*

Chemophobia can be seen in a pronounced manner within the realms of beauty, nutrition, and general health. For example, a common sentiment exists which can be conveyed by this seemingly logical mantra:

*If you can't pronounce it, don't eat it!*

This mantra, peddled by "Food Babe" and other such public figures, is clearly directed toward nutrition labels that list all of the ingredients in a food product, some of which are quite unfamiliar to most people. If this advice were actually followed, it would be deadly for the majority of us, at least certainly for those without training in chemical nomenclature. The names of many essential vitamins and nutrients are difficult to pronounce, because chemists name them without the general public in mind. Try and pronounce pyridoxine, which is vitamin B6, or cobalamin, which is vitamin B12, or cholecalciferol, which is vitamin D3. There is no correlation between how many Z's or X's there are in a name and how harmful that molecule is to consume. There is no relationship between the number of syllables in a name and the bioactivity of the corresponding molecule. These are just names, and, as Shakespeare asked, what's in a name?

The claim "chemical-free" in reference to any product is patently absurd.

Furthermore, there are those that sell or recommend food products who have the audacity to claim that they are "chemical-free." This is completely absurd, as every single thing you have ever eaten is literally nothing but chemicals. This is because anything that is made of the chemical elements qualifies as a chemical. The only thing we can see with our eyes that could be accurately described as "chemical-free" is light, and that's not exactly going to fulfill any of your dietary requirements. A less blatantly false claim to make is that of a product being free of some specific chemical. This may or may not have legitimacy, it depends on what is being referenced. Sometimes certain chemicals are identified as harmful and subsequently removed from a particular manufacturing process, and this could be rightfully advertised and applauded. But more often than not, this claim is simply a marketing tactic which is exploitative of chemophobia. It references something that was never present in the product, or which is not inherently harmful in the first place, and thus does not convey any important information about the product. Perhaps a particular shampoo claims to contain half as many chemicals as another brand. So what? Which chemicals? Why does it matter how many chemicals there are? Isn't one harmful chemical immeasurably worse than a thousand benign ones?

Sometimes the blame doesn't lie with the manufacturer. Often an activist group will catch wind of a compound found in, say, a laundry detergent, which is categorized as a **carcinogen**, meaning a cancer-causing agent. Without considering the concentrations listed, they use their outrage to fuel an attack on the manufacturer, blissfully unaware that carcinogenicity, like toxicity, is inextricably tied to the dosage. Not knowing that the amount present in the product poses no danger whatsoever, they enact a smear campaign. In response, the manufacturer goes out of their way, wasting money to try and remove a harmless component from their product, simply to avoid bad press. Victorious and regarding themselves heroic, the activists pat themselves on the back on the way to their next book club meeting. And yet, if carcinogens in household products is their pet cause, where is the outrage toward Mother Nature and all of her carcinogens? Why get angry about formaldehyde being present in trace amounts in laundry detergent, but ignore the naturally occurring formaldehyde present in a thousand times greater concentration in apples, pears, and potatoes? Where is the anti-earth lobby and the ensuing boycott against pretty much everything that comes out of the ground?

Alt-health websites will go one step further, as they will address not what actually exists in products, but will rather pull ingredients out of thin air, sometimes going so far as to make up names for chemicals that don't exist. They might even complain about harmful effects that occur when a product is ingested, neglecting to consider that something like laundry detergent is obviously not intended to be ingested. Quite commonly, such websites will offer their own alternative products for sale, which rather predictably, claim to be free of all such "nasty chemicals," though they are often full of other ones.

Whether we are talking about duplicitous marketing tactics, or pseudo-woke outrage from consumers, the common theme is chemophobia. The public displays chemophobia in myriad ways. And when an advertisement exploits this chemophobia, it is nothing more than an appeal to the fear of the unfamiliar. It is a gentle but empty reassurance. A whisper in your ear, coaxing you to the conclusion that a product is not just safe to buy, but ethical to buy. In buying their product you are supporting a company that did the right thing. The ad sells you your own virtuous self-image at a tidy markup.

> **TL;DR—Exploitation of chemophobia is a common and effective marketing tactic, though fundamentally deceptive.**

Many of us have a tendency to think that we are immune to advertising. But the truth is that none of us are. Advertising works. If it didn't, no one would do it, as it's extremely expensive. Every time you see a product being marketed a certain way, subliminally suggesting that you associate some product with healthfulness, or deliciousness, or masculinity, or femininity, or luxury, you are being manipulated. But when a brand tries to get you to associate a particular deodorant with "manliness," there isn't really any broader harm being done, provided that the product is safe to use. If it works on you, and you are slightly more likely to purchase that product, then the marketing campaign was a success, and they will have recouped their investment. It's just the battlefield of capitalism. Some brands do better than others, and it's essentially all the same to the rest of us. But the alt-health industry is not so innocent. When scientific principles are misrepresented, and when chemophobia is promoted, society suffers. It causes people to spend a fortune on "natural" household products instead of their more economical and equivalently safe counterparts. It leads to people eschewing legitimate medical treatment in favor of snake oil. Beyond these immediately tangible ramifications, it also fundamentally affects the way the public perceives the scientific community. It sows the seeds of distrust and manifests false divides that must then be dealt with rather than tackling the real issues. Because of all the misdirected energy, and the flared emotions involved in these skirmishes, it then becomes that much more difficult for the population to become sufficiently informed, as it is ten times harder to un-scare a person than it is to scare them. How can we know when we are being manipulated in this manner? How do we spot this deceptive activity, and what knowledge do we require in order to do so? That's precisely what we are here to answer, so let's move forward and dig a little deeper.

CHAPTER 4

# The Molecules of Life

By this point, we know a few things about molecules. But most of us are not so interested in just any molecules. We are not all chemists. We don't regularly think about rare and exotic molecules, or completely hypothetical molecules. Most of us think only about the molecules that are relevant to all of us in our everyday lives. We think about the molecules inside our bodies, and the molecules that come into contact with our bodies, whether through consumption, inhalation, absorption, or other such physical and chemical processes. In short, we are primarily concerned with our own general health, and the chemistry involved with its maintenance. For this reason, from the more general discussion of molecules we have engaged in thus far, we will now narrow our focus toward the molecules of life. We already mentioned that just a handful of elements make up most of what's inside living organisms, those being carbon, hydrogen, oxygen, nitrogen, phosphorus, and sulfur. But what kinds of molecules do these elements come together to form, so as to comprise and sustain this incredibly complex dance of consciousness that we call a human being? What's inside of us, exactly?

Apart from lots and lots of water, our bodies contain four main classes of biomolecules. These are proteins, nucleic acids, lipids, and carbohydrates. You most likely have heard of these before, but we have to go beyond hearsay and factoids. If we want to understand general health and medicine, we are going to have to become familiar with the precise structure and function of these things. To begin, let's make it a little easier on ourselves. Lipids and carbohydrates can wait for a few chapters, as they will be more relevant when we talk about cells and cellular activity. **Cells** are the basic structural and functional unit of life, at least as we know it, but each cell is made of millions of different molecules, so we have some work to do before we get there. We first have to talk about biomolecules, beginning with proteins and nucleic acids. This may again seem like homework, but remember how painless Chapter One was? Remember how satisfying it was to look at that molecule in line notation and actually comprehend what it represents? Well let me dangle another carrot before you.

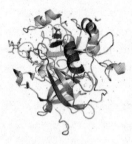

A visual representation of a protein.

Here we see a representation of a protein. Apart from water, proteins make up most of your body. So what is this big blob? What do the different sections of the blob represent? What happened to the line notation we spent so much time learning how to interpret? Just as with the small molecules from before, after some brief explanations, this too will become intelligible, I promise.

So what is a protein? Like most other large biomolecules, **proteins** are polymers. This means that they consist of many, many repeating units called monomers, linked in linear fashion, like beads on a string. Poly means many, and mono means one, so many monomers linked together makes a polymer.

A polymer consists of many monomers linked in succession.

Each biological polymer, or biopolymer, has a different kind of monomer, and proteins are made of monomers called amino acids. Amino acids have the general structure shown below. The part with the nitrogen atom is a functional group called an amino group, which is found in molecules called amines. The part with the oxygen atoms is called a carboxyl group, which is found in molecules called carboxylic acids. Putting them together, amino plus acid equals amino acid. On the carbon between them, there is something called an "R group," which is a side chain of varying identity, depending on which amino acid we are looking at.

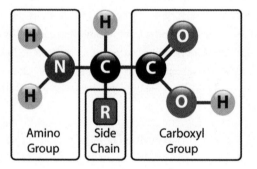

The general structure of an amino acid.

If this R group is just a hydrogen atom, we get an amino acid called glycine. If it is $CH_3$, which is called a methyl group, we get alanine. There are twenty different amino acids in living organisms, and they possess side chains with dramatically different properties. Some are small and some are bulky. Some interact well with water and some don't. Some act as acids and some as bases, which are terms that describe how well a functional group is able to donate or accept a hydrogen ion, an example of extremely simple and commonplace chemistry.

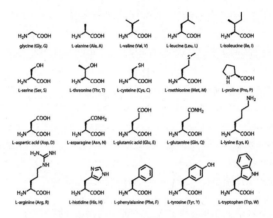

The twenty amino acids present in biological systems.

Amino acids polymerize, meaning they join together like those beads on a string, and when a couple hundred or more of these amino acids are combined in succession, we get a protein. Because of the practically endless ways that twenty different amino acids can combine, the number of possible proteins is essentially limitless.

*TL;DR—Proteins are polymers made of twenty different amino acids in some sequence.*

The key thing to understand about proteins is that their function in the body is determined by their shape. Proteins are long, linear molecules, but they fold up in highly specific ways so that different functional groups can interact with one another. The more interactions it can make, between positive and negative charges for example, the more stable it becomes, and this folding pattern is entirely dependent on the specific sequence of amino acids. If a protein were to have its sequence of amino acids altered, it would likely alter the folding pattern to some degree, often dramatically, so as to again maximize the electrostatic interactions it is participating in. It's a bit like a tangled string of Christmas lights with twenty different colors, if there were specific attractions between the different colors that caused it to tangle in a specific, non-random way.

Some proteins end up being purely structural, holding the body together, but some proteins actually do chemistry. Proteins that facilitate chemical reactions in the body are called enzymes. These are nature's solution to the problem of doing complex chemistry in such a limited environment. Nature can only use water as a solvent. It can't modify the temperature, or pressure, or pH. It can't regularly utilize rare elements like transition metals. It can't use any reagents that would destroy biological organisms. Without the ability to use extreme conditions, biochemical reactions are achieved at body temperature in what is called the active site of an enzyme, which is like a little pocket in the protein where molecules enter.

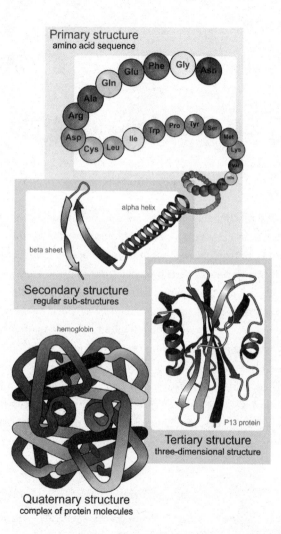

Levels of protein structure.

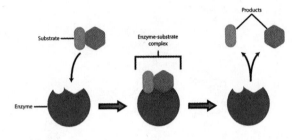

An enzyme operating on its substrate.

The local functionality of the active site, determined by the specific amino acids present in that region, is what allows for chemistry to take place. It might have a shape that encourages a molecule to bend a certain way in order to make more interactions with the enzyme, and in doing so, a particular bond in that molecule is contorted and weakened, making the molecule easier to break in half. It might enhance the reactivity of a particular functional group on the molecule, by influencing the distribution of electron density. Every enzyme has a specific molecule, or substrate, that it acts upon, a specific reaction that it can then catalyze, and a highly specific mechanism by which this reaction is achieved. Beyond enzymes, there are several other types of proteins, and some of these will be introduced a bit later.

*TL;DR—Proteins that facilitate specific chemical reactions are called enzymes.*

In terms of interpreting the blob we looked at earlier, it is important to understand that proteins are so large, and contain so many atoms, that it is impractical to use conventional line notation, as this would involve representing thousands of individual atoms. Instead, there are other ways of depicting proteins. We can use colored strings and strips and arrows to illustrate the way the backbone snakes around, forming winding alpha helices, hairpin turns, and beta-pleated sheets, where the strand doubles back on itself to run parallel with a previous segment. We can also use more of a space-filling model so that we can see what all the side chains are doing. But in short, as long as we understand how to use line notation to represent a few consecutive amino acids from this long sequence, we can easily do so for any such section, allowing us to zoom in on key areas as needed, such as the active site of an enzyme. This structural knowledge is gained through sophisticated techniques, like X-ray crystallography, which allows us to see the three-dimensional structure of a protein, as well as computer simulations, which allow us to predict the structure of a protein simply by introducing an amino acid sequence and applying the laws of physics. And just like that, this blob now makes some sense, and we have begun to bridge our understanding from the realm of small molecules to the realm of large biomolecules.

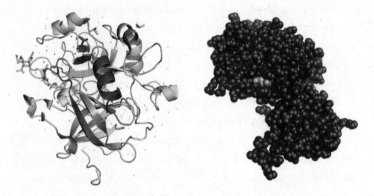

Various ways of visually representing proteins.

But we can't stop now, proteins are only one type of biopolymer. We also want to understand nucleic acids, such as **DNA**, which is short for deoxyribonucleic acid. As pretty much everyone has likely heard since elementary school, this molecule serves as the genetic code. But what does that mean? How does it encode information, and how is that information transmitted? This will take a moment to explain, so let's start by looking at the structure of DNA, beginning with the monomers it's made of. These are called nucleotides. Unlike the twenty amino acids, there are only four nucleotides, each of which has three sections. The cyclic part in the middle is called a sugar. The part with the phosphorus atom is called a phosphate ester. And the remaining part is called a nitrogenous base, or usually, simply a base. The base is the only part that varies, changing its identity amongst four possibilities, which are named adenine, thymine, guanine, and cytosine, and abbreviated as A, T, G, or C.

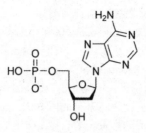

The structure of a nucleotide.

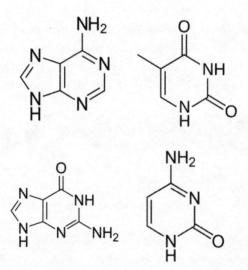

DNA bases adenine (A), thymine (T), guanine (G), and cytosine (C).

Nucleotides are the monomers that make up nucleic acids, which means they must polymerize, and they do so at the

phosphate groups. This happens many millions of times
to produce incredibly long molecules. Beyond this, a DNA
molecule is also double stranded. This means that there are
two long strands of nucleotides running alongside one another,
each of which has its bases pointing toward the inside, and the
rest of each strand, which we call the backbone, twists along
the outside in helical fashion. This is why we say that DNA is a
double helix. So while some generalized image of DNA looking
like a twisted ladder was likely already quite familiar to you,
we now understand the structure of this molecule at the most
fundamental level. We know that DNA is comprised of two
strands of nucleotides, and we know precisely what those are,
atom for atom, just as we can now zoom in on any portion of a
protein and discuss its amino acid sequence.

Cytosine        Guanine        Adenine        Thymine

Deoxyribonucleic acid, or DNA.

Regarding the bases, it is the case that each one pairs with a
specific base on the other side, A with T, and G with C. Two
of the bases have two rings while the other two have just one,
and only when combining one of each will a pair exhibit the
appropriate width. So purely by size, A will not pair with G, and
C will not pair with T. Beyond this, the bases have functional
groups that interact with each other when paired appropriately.

Adenine makes precisely two favorable electrostatic interactions with thymine, which are called hydrogen bonds, and guanine makes three such interactions with cytosine. Just as with any other interactions of this nature, these bring the system to a lower energy, or a more stable state, so this process of so-called base pairing is highly specific, making the two strands of any DNA molecule precise complements of one another. The sequence of one strand necessarily determines the sequence of the other.

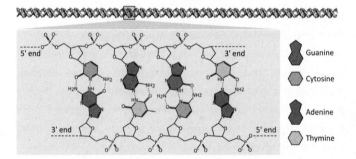

Base pairing in DNA is highly specific, with hydrogen bonds between the bases represented by dashed lines.

*TL;DR—DNA is a polymer consisting of two strands twisting together in a double helix, each made of four different nucleotides in a complementary sequence.*

This fact is also the key to understanding how DNA encodes information. We are now ready to tackle the concept of **gene expression**. It will be a bit of a challenge, but it is absolutely imperative. Gene expression is the key to understanding genetics, evolution, human development, disease, and much more. This singular concept, more so than any other, is the key to understanding biology. So let's take a breath and dig right in.

Your body is made of cells, and a particular region of each cell, which is called the nucleus, contains a copy of all of your DNA. This is not to be confused with the nucleus of an atom that we are familiar with from our discussion of chemistry, it's just a coincidence that we use this word in these two separate contexts. In biology, the nucleus is the central compartment of the cell that contains the genetic material. This genetic material is packaged in structures called **chromosomes**, whereby DNA molecules that are many millions of base pairs long wrap around proteins called histones to save space, as there is so much DNA to store. That should offer some new perspective on an image of a chromosome, which is just as familiar as any image of DNA portrayed as a ladder, and just as commonly misunderstood.

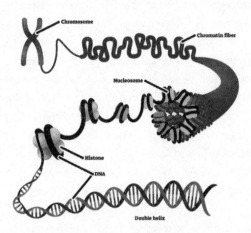

The structure of a chromosome.

The amount of DNA in just one chromosome is truly astounding, and humans have twenty-three pairs of them, for a total of forty-six in each cell, amounting to about three billion base pairs in total. If we were to take all of the DNA from every cell in just one human being and stretch it out in linear fashion, we could get to the edge of the solar system and back! This provides some idea of how important it is for DNA to be packaged in a compact way within chromosomes.

*TL;DR—DNA is stored in compact structures called chromosomes.*

Just as fascinating is the fact that DNA is over 99 percent identical across all human beings. And yet the fraction of a percent that differs is enough to produce all of the physical variety amongst our species. How are these differences manifested? Well, certain sections of a chromosome are called genes. These sections, which comprise 3 percent of the genetic material, code for the production of proteins, which as we said, make up most of what you are. This happens via two sequential processes called transcription and translation. Let's tackle transcription first. An enzyme called RNA polymerase will bind to a specific segment of DNA because it recognizes some highly specific sequence of bases, making favorable interactions with that sequence in particular. Be careful to understand that when we say "recognize" we do not imply sentience or intent. It is simply that the shape and composition of the enzyme is such that it makes enough favorable interactions with that specific sequence of bases that it is thermodynamically favorable for it to bind at that location. The stickiness of plus and minus strikes again. Then, due to conformational changes induced by the act of binding, meaning changes in shape, the enzyme is able to pry apart the strands of DNA and move along one of the strands, assembling a complementary strand made out of RNA, or ribonucleic acid, which is almost the same as DNA. This follows the same rules for base pairing as the complementary strands of DNA, the main difference is that instead of thymine, or T, RNA uses a slightly different base called uracil, or U. Once RNA polymerase is done reading the gene, the DNA goes back to normal, but something called a messenger RNA, or mRNA, has been produced, which encodes the information in the gene. So a gene will contain a specific sequence of bases, which therefore codes for a specific mRNA, according to the same rules of base pairing that apply to DNA.

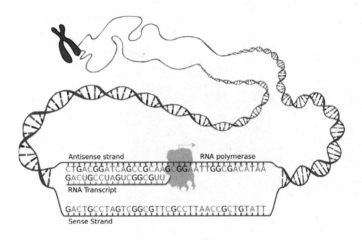

During transcription, the antisense strand of a gene is used by RNA polymerase as a template to generate a messenger RNA, or mRNA, shown in green.

After some modifications are made, this mRNA will then leave the central nucleus and find something called a ribosome, where translation occurs. In translation, the mRNA is read by the ribosome in groups of three bases, each of which is called a codon. Other molecules, which we can refer to as transfer RNA, or tRNA, then enter the ribosome and bind to the codons. Each tRNA has an anticodon on one end, which acts as a complement to a specific codon, and on the other end there is a specific amino acid. So a particular sequence of three bases

on the mRNA will allow only one particular tRNA to adhere, and
thus codes for the particular amino acid that is associated with
that tRNA.

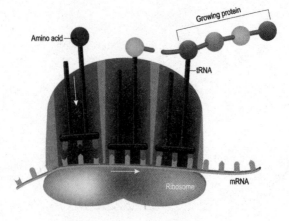

During translation, the mRNA is used as a template
for protein synthesis within a ribosome.

With three bases per codon, and four possibilities per base,
we have $4^3 = 64$ possible codons. There are only twenty
amino acids to code for, so multiple codons will code for the
same amino acid, and then certain special codons act as start
and stop codons, which initiate and terminate translation,
respectively.

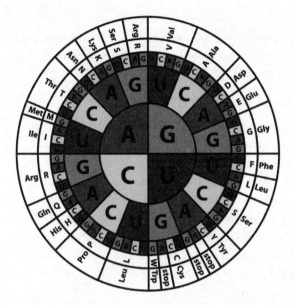

Each series of three bases on the mRNA acts as a codon, which codes for a specific tRNA, and therefore a specific amino acid in the resulting protein.

With one tRNA and its associated amino acid already bound to the mRNA, when the next tRNA arrives to bind with the next codon, the two amino acids on the adjacent tRNA molecules react and become connected. This happens all the way down the mRNA, slowly growing the amino acid chain. When finished, everything disassembles, the protein folds up according to its amino acid sequence in a nearby part of the cell and is now ready to serve its function.

> *TL;DR—Gene expression involves the production of a specific protein according to the sequence of DNA bases found within a gene, via transcription and translation.*

So to summarize, in transcription, the DNA bases within a gene serve as a template for the production of an mRNA molecule by the enzyme RNA polymerase. Then in translation, the mRNA serves as a template for the production of a protein by a ribosome and many tRNA molecules. Our DNA has many thousands of different genes that code for all the different proteins in our bodies, and these are expressed at specific times when appropriate signals are received. From DNA to protein, that's gene expression, which is the central theme of molecular biology.

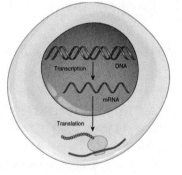

Gene expression involves the production of a protein
via transcription and translation.

Obviously, this process has been simplified dramatically. Many details have been omitted, as a textbook of a thousand pages could be filled with what we took only a few paragraphs to describe. For this reason, it will be tempting for most to interpret the whole show as sentient little factory workers building molecules with purpose and resolve. But it can't be stressed enough that this is all completely spontaneous, occurring due to chemical principles, and propelled by other energy-producing reactions in the cell that we will discuss later. Biochemistry, no matter how complex, is driven strictly by interactions between positive and negative charges that are energetically favorable. These are the same interactions between plus and minus that are responsible for the existence of atoms, their assembly into molecules, and the properties that cause molecules to react with one another. Just one fundamental force, the electromagnetic force, allows for this amazing molecular dance, which results in the production of all the proteins that give rise to your form during embryonic development, and maintain that form during the human lifetime. While there is so much more that could be said about rudimentary biochemical processes, we will have to keep our eyes on the prize, and that is an ability to discern between scientific fact and misinformation, so let's now direct our efforts toward understanding general health.

# CHAPTER 5

# The Molecular Basis of Wellness

Living organisms are concerned with self-preservation. It's instinctual. We humans are afraid of the dark, we are afraid of falling, we seek to reproduce, and we don't want to die. There is little else that could be considered so innate, so central to the human psyche, or that of any other animal for that matter. So how do living organisms avoid death? Given that around the biosphere, the most common cause of expiration is getting consumed by something else, plants and molds and even some animals produce poisonous compounds to discourage other organisms from making a meal of them. Mammals tend not to have such chemical defenses, so getting eaten has historically been of great concern for us. But *Homo sapiens* have built a civilization as a safety net. Getting mauled by a lion in the savanna is no longer an especially probable demise. Some people are involved in tragic accidents, but most of us die because we get very old or very sick. Cancer, lung disease, heart disease, stroke, and diabetes are the biggest killers in modern times. How can we avoid these means of expiration? How can we stay healthy? This question has had many answers over the millennia, but only since the development of modern chemistry and biochemistry have we had even a sliver of a hope of answering it. One need only glance at a book of medieval medical practices to thank one's lucky stars for living in the twenty-first century. It's a wonder that mankind has survived until this point at all. Fortunately, we now operate under a superior framework.

We are but molecules. In fact, everything we interact with is but molecules. Anything we see in our macroscopic reality is the consequence of events on a submicroscopic level. The sun shines because of the fusion of atomic nuclei. Hurricanes and other huge storms are the result of the behavior of molecules in the atmosphere. The biological world is not exempt from this fact. Whether we are trying to understand why a volcano erupts or how a flower blooms, we turn to the realm of molecules.

Our own bodies are no exception. All that we perceive and all that we do is possible due to the existence of, and interactions between, all the molecules that comprise our bodies, and all the molecules we exchange with our environment. This was not known in times past, which left us stabbing in the dark as to how one can best treat illness, turning almost exclusively to pseudoscience or even sorcery in absence of any real physical knowledge. Now we understand that all disease has a molecular basis. In short, illness occurs because molecules are doing things they aren't supposed to do, or they aren't doing things they are supposed to do, whether due to genetics or faulty molecular design. We will try to expand on this vague and overly reductive statement when appropriate over the upcoming chapters without drifting into a medical lecture, but generally speaking, it is the first and most fundamental notion that we must accept. Anyone that discusses any kind of medical treatment without at least alluding to what is happening on the molecular level is either not legitimately knowledgeable or behaving deceptively. Often, both.

> **TL;DR—All disease has a molecular basis.**

Let's go through an example, which will now be possible with our newfound understanding of gene expression. Let's say that the DNA in a cell sustains a **mutation**, meaning that one or more of the bases undergo some chemical change. This can happen for a number of reasons. It could occur because of exposure to a mutagen, which may take the form of a harmful chemical, as most would immediately imagine, but this could also be ultraviolet light from the sun. Alteration of bases even happens by accident during DNA replication, where an enzyme makes a mistake and puts the wrong base in a particular location. Whatever the case may be, in humans there are enzymes which are responsible for recognizing these discrepancies and fixing them. But if they miss one, which is uncommon but does happen from time to time, this will cause a problem during the next round of DNA replication. We will discuss this process in more detail later, but for now we simply have to know that this is the method by which all of your DNA is copied, one strand at a time, to produce two complete sets of DNA prior to cell division.

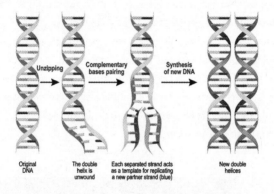

In DNA replication, each strand acts as the template for its new complementary strand.

If there is an error in the template strand, the enzyme called DNA polymerase, which is responsible for reading the template strand and synthesizing the new complementary strand, will read the mutated base and typically implement the wrong base as its complement at that particular location. From then on, the genome in that cell is permanently altered, along with all of its daughter cells, which are the new cells that form when a cell divides. In the case that the alteration involves only a single base pair, we would call this a point mutation.

Let's now say that this point mutation occurred within a
gene. This would change one codon in the mRNA produced
during transcription, and thus could potentially change the
corresponding tRNA and the associated amino acid for that
position during translation. This will change the resulting
protein, sometimes in a subtle way, but sometimes in a dramatic
way. It could cause the protein to lose its function, either
partially or entirely. It could cause the protein to adopt a new,
harmful function. Either way, disease may result, particularly if
such a mutation exists in both of the chromosomes that contain
this gene, and there are many examples of diseases that arise in
precisely this way, such as cystic fibrosis, sickle-cell anemia, and
even some kinds of cancer.

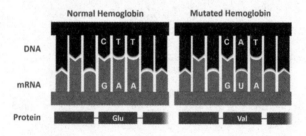

Sickle-cell anemia is the result of a point mutation in the gene that encodes
hemoglobin, an important protein found in red blood cells.

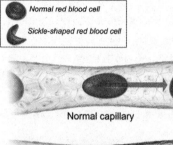

**Normal capillary**

**Sickle Cell Anemia**

The mutated hemoglobin causes red blood cells to take on a sickle-like shape, which then clog blood vessels.

This is just one relatively simple example of what can go wrong on the molecular level. When we can identify such a specific cause for a disease, whether a point mutation, some other genetic aberration, or a related situation, any outlet that describes treatment without discussing this fundamental cause on the molecular level is either misinformed or deliberately manipulative. There is nothing about exercise or nutrition that could ever repair something like genetic mutation or address its immediate effects. At best, such practices can delay the onset of certain diseases that arise from these genetic predispositions, which is beneficial but not a permanent solution. The only recourse is to address the problem on the molecular level. This could involve a drug that fits into the active site of a receptor protein, which would thereby be capable of either promoting or disrupting cell communication. Or it could also be achieved through methods at the forefront of biotechnology. This is how modern science-based medicine works, which means this is how Western medicine works. It is medicine that is derived from strategies built upon the knowledge produced by the scientific community. There are treatments that currently exist for so many conditions that utilize strategies of this nature and help millions of people function normally.

> **TL;DR—Western medicine is science-based medicine.**

Let's recall the list of big killers at the beginning of the chapter. Cancer, lung disease, heart disease, stroke, diabetes, what does Western medicine do with these? Cancer is a meaty one, which we will dissect a bit later. But take something like heart disease. What is the approach taken by conventional heart medication? Well depending on the specific condition, many approaches can be utilized. Anticoagulants, for example, decrease blood clotting by blocking the production or activation of proteins in the liver that assist in the formation of blood clots, thereby minimizing the potential for strokes and heart attacks. Angiotensin-converting enzyme inhibitors, or ACE inhibitors, inhibit the enzymatic activity that converts a molecule called angiotensin I into angiotensin II, which is involved in blood vessel constriction. Blocking its production prevents the narrowing of blood vessels, making it easier for blood to flow through, thereby reducing strain on the heart. Beta-adrenergic blocking agents, or beta blockers, help treat an irregular heartbeat by inhibiting adrenaline receptors, thus blocking the activity of signaling molecules that would cause the heart to beat faster and more forcefully.

There are about a dozen other approaches to manipulating the heart and the rest of the circulatory system, and each of them exploits a unique physiological aspect of circulation, which is only possible through an intimate understanding of human biology and anatomy. This is what medicine looks like when it operates on the molecular level. It's true that lifestyle choices may have gotten someone into a particular heart-related predicament, such as bad dietary habits and lack of exercise, but this is frequently not the case. Once again, there is often simply a genetic predisposition toward certain conditions, that may or may not be mitigated by diet and exercise, which inevitably becomes a problem, if they are not already problematic from birth. But all of the above is difficult to understand, which explains why so many disregard it entirely rather than attempting to understand it.

There is no shortage of criticism regarding Western medicine on the internet, and the vast majority of it fits into the two categories we described previously. There are those who are simply misinformed, and do not comprehend the biochemical principles surrounding illness, but innocently regurgitate misinformation because it "feels" right to them, for whatever reason. And then there are those who seek to manipulate by knowingly spouting misinformation. The alt-health industry, which peddles alternative health solutions, exploits public distrust of Western medicine to sell products. To be clear, this is not an indictment on those who criticize the ethics of large industries, at times rightfully, of which medicine and health insurance both qualify. This is about disease and treatment. This is about science. A popular narrative exists, which states that Western medicine, namely pharmaceutical drugs, treat the symptoms of disease and not the cause. Conversely, alternative and "holistic" medicines treat the real cause. In actuality, this is the precise opposite of the truth. As stated, all disease has a molecular basis, and science-based medicine specifically addresses that molecular basis. If a genetic abnormality results in some mutant protein with deleterious results, that protein is the cause, and only a drug that is developed with this cause in mind can act as the solution. Anything else produces a placebo effect at best. We will examine specific examples to help overturn this false narrative over the coming chapters.

> *TL;DR—Anyone who says that Western medicine treats the symptoms while alternative medicine treats the cause has it precisely backwards.*

Next, let's continue a thread from an earlier chapter. We discussed the definitions of the words "natural" and "synthetic," and how they are misunderstood. We are now ready to examine a specific exploitation of this confusion in a medicinal context. Most people know that we need to consume things called vitamins. What are these exactly? In short, a **vitamin** is a compound that the body needs to have present in some concentration in order to function properly, and this is typically a compound that the body can't make on its own. Many molecules needed by the body are made inside the body. Over the course of biological evolution, the capacity to synthesize certain compounds has been lost due to random mutation, which did not impact organisms as negatively as one would expect, because those compounds happened to be abundant in their food sources and thus always present at necessary levels in every cell. What this spells for us is that we absolutely must continue eating food that contains those compounds, or we get into trouble.

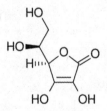

Ascorbic acid, also known as vitamin C.

Take vitamin C for example. This is the vitamin nickname for a compound referred to by chemists as ascorbic acid. Among other things, this compound plays a vital role in the synthesis of collagen, which is a structural protein, and the single most abundant protein in the body. The enzymes that synthesize collagen require vitamin C to be present, to act as something called a cofactor, in order to function properly. Without this compound, the result is vitamin C deficiency, also known as scurvy, whereby collagen synthesis is impaired, resulting in a variety of debilitating symptoms. This disease was famous among sailors of the sixteenth to eighteenth centuries, since fruits and vegetables containing vitamin C would tend to spoil over long journeys at sea, and after a month or so of the deficiency, symptoms developed, and many sailors would die. It wasn't until the late eighteenth century that the connection was made between citrus fruits and the ailment, such that lemon juice could be stored and administered as a preventive measure. So clearly, vitamin C consumption is crucial.

Fast forward to today. Vitamin supplements are ubiquitous, and it is difficult for suppliers to stand out from the pack. Marketing schemes must be devised, and I'd like to share one that I've stumbled upon. Of course, we know that vitamin C is found in nature, but we can also synthesize it in the lab, just like any other natural product. In fact, this is often quite a bit cheaper than extraction and isolation from food sources, and typically results in higher purity as well. It remains true that fulfilling all your nutritional requirements directly through your diet is a great idea, but this is not possible for everyone in the world, due to a number of factors, like poverty or climate. Not all the people have access to all the foods, and some have access but can't afford them. Supplements are useful for fulfilling any requirements that are not met through the consumption of food for any reason, and with vitamin C, or ascorbic acid, it does not matter whether this is natural or synthetic. Regardless of origin, the molecule will interact with the collagen-producing enzymes in the same way and enable collagen synthesis. Just as we already learned, no matter how the molecule is built, the structure is identical, so the bioactivity is identical.

But some spin a different tale. Some speak of a "vitamin C complex." This is some entity, with a structure that is not made explicitly clear, that is required for optimal biological function. Circular diagrams are displayed with various letters in quadrants, such as J, P, and K, and sometimes the word copper, all circumscribed by a shell that is labeled ascorbic acid. The claim being made is that synthetic vitamin C is just ascorbic acid, which is only the shell of the entire complex, and is therefore insufficient in maintaining your health. Claims are made that ascorbic acid is synthesized from coal tar or other scary-sounding starting materials and is also full of "additives," or other miscellanea. They go on to insist that people should beware of those who sell synthetic ascorbic acid, as it just isn't the real deal, and that they are quite fortunate to have stumbled upon this very important PSA.

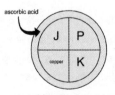

A visual representation of the alleged "vitamin C complex."

There is a lot to unpack here. Let's talk about the claim first, before examining the people making the claim. First, ascorbic acid and vitamin C are synonymous. They are simply the chemical name and vitamin nickname for the same compound. Molecules often have multiple names, that's just how it goes. Chicago, the Windy City. Ascorbic acid, vitamin C. Same city, same molecule. So right off the bat, this claim that there is some discrepancy is objectively false, and indicative of duplicitous intent.

Second, what can this diagram possibly mean? So many things are unclear. What are these quadrants? Why are there no molecular structures anywhere? What is this "shell" and how does it work? How can ascorbic acid act as a shell? You've seen the structure; does it polymerize somehow? How does that work, and why isn't that shown more explicitly? Why would it matter what the starting material is for ascorbic acid synthesis? Coal is carbon, and all organic molecules have carbon atoms in them. It wouldn't mean that there is coal in the vitamin, even if the claim was true about coal tar, which it isn't. What do these other components do? Are they cofactors as well? Which enzymes do they interact with? Is there evidence to reference?

None of it holds up to an ounce of scrutiny, and its fabricators know that very few people will make any effort to scrutinize. Because in absence of scientific knowledge, particularly for those with an anti-establishment bias, it sounds right. It feels right. Man, in his limitless hubris, made a thing to imitate nature, and those same dishonest people want to sell it to you. But don't worry, here are some heroes speaking out against the man, telling the truth about the nasty chemicals and offering the warm, fuzzy, all-natural version. It doesn't automatically register with us that such people are always trying to sell you something. They often have "Dr." in their title, they wear a white lab coat for no apparent reason, and they speak very confidently using scientific-sounding terminology. How could they be lying?

*TL;DR—People lie on the internet. A lot.*

Suddenly we find ourselves at a crossroads. I mean us. You, the reader, and me, the writer. I may not be selling anything, as you've already purchased this book. But some of you may be thinking, "Why should I believe this guy? What if he's the one that's lying, and not the vitamin C complex folks?" Fair enough. It is frustrating to encounter contradictory information on the internet, or in books for that matter, and I don't wish to diminish this sentiment, as it is a universal feeling, whether one lacks knowledge in science, or politics, or any other field. This is an instance where I must then offer you another strategy to determine for yourself who is being truthful.

The first strategy we outlined is to learn biochemistry, and we did put a dent in that, but most of us don't have time to get too far with this endeavor. Furthermore, the bit of biochemistry we learned together wasn't quite enough to tell us what to think in this particular instance. So we turn to Google. I know that sounds dangerous, given the mountains of misinformation, but we have to be clever. Google is a phenomenal tool if you use it properly. In this case, the claim being made is that ascorbic acid alone doesn't do the trick, and that we need the entire vitamin C complex. Let's do something that should seem obvious, but that most people fail to consider. Simply Google the term "vitamin C complex." If you're near a computer or have your phone on you, I encourage you to do this in real time as an exercise. But don't hit enter just yet. First, make some predictions. If the vitamin C complex exists, what should we expect to find? Well, we probably ought to turn up a few hits of a scholarly nature. If this complex is documented and understood by the scientific community, then somebody must be studying it, and we should find a neutral source outlining some relevant details. This could be an academic journal article, it might be a university website for biochemistry students, or at least a blog post or two. Even if very little content can be found, there must be a Wikipedia page for this thing, no? Essentially anything that is a thing is on Wikipedia, right? Ok, that's enough predicting. Hit enter. What do you get? Any of the above? Nope. Not a single one. Page after page of products. Just a bunch of people trying to sell you stuff.

Even if you were to have zero knowledge of biochemistry, the results of this Google search should convince you that there is no such thing as the vitamin C complex. It is as hollow as the feeling evoked by the false claim that ascorbic acid is a "shell" of something greater. The whole concept is a fabrication, a false sense of urgency, a convenient solution to a problem you don't have, in the form of something that does not exist. In short, it is nothing but a marketing tactic. It is a way of manipulating you into purchasing something, and the narrative employed couldn't be more transparent. Man-made, bad. Industry, bad. Nature, good. Our product, with the sun and trees on the bottle, good. It's so infantile that we should all feel insulted. But so many of us do not apply even this minimal level of scrutiny in our daily lives, least of all for our health, which is where it should be applied the most.

> **TL;DR—Google searches can tell you if something is real or not, just be clever.**

Although the internet is the entity that sold you this lie, the internet can also be your friend in the endeavor for truth, because manipulation and lies, which suit the very few, are always outnumbered by facts, which suit the masses. We fall for these types of hoaxes almost exclusively due to confirmation bias, which undermines the logical impulse to compare information with other sources. If we want something to be true, being told that it is true feels so good that our capacity for rational analysis shuts down. Salesmen know this, and they know it well, just as they have since the dawn of commerce. In this particular case, the result is not so devastating. If you waste twenty extra bucks on an expensive bottle of vitamin C supplements that are identical to the cheap, generic bottle, it's not the end of the world. But we will later outline examples whereby this precise tactic encourages people to avoid legitimate medical treatment in favor of unsubstantiated alternatives, which can frequently result in serious health consequences, or even death. That is the type of urgency we are building toward, and that is the main purpose of this book. Let's move forward and dig a little deeper.

CHAPTER 6

# The Rise of the Alt-Health Industry

For decades now, human civilization has been following a path that is untenable. We have been stripping the earth of coal, oil, and natural gas, and using these resources to fuel a society that is growing at breakneck speed, with insufficient regard for any and all environmental ramifications. These practices are perhaps the most likely origin for the public distrust of industry. This distrust is precisely why Science with a capital S makes so many people think of billowing smokestacks, polar bears balancing precariously on melting blocks of ice, and barrels full of toxic green goo. The nature of capitalism is such that this activity is exclusively profit-driven, and the rest of us suffer the consequences of deregulation and loose ethics. They reap the profits, and we pay for the environmental and social costs of their activities.

But then, like a glistening yet renewable beacon of hope, alternative energy will rise from the tar pits to save the day.

Wind, solar, and other innovations in energy production promise to reduce our reliance on fossil fuels and bypass the greed of the industries that produce them, saving humanity in the process. That's what alternative means, it represents the notion that another possibility exists. In this case a better one, a more ethical one. Now doesn't that instill you with all kinds of zeal and happy feelings? Swim around in that warmth for a moment. Bask in the radiance of the alternative.

Solar and wind energy are examples of alternative energy.

This positive connotation for the word "alternative" is pivotal in understanding public perception of alternative medicine. Most people would argue that, just in the way that alternative energy bypasses the greed of fossil fuel industries and saves our species, alternative medicine bypasses the greed of the pharmaceutical industry and saves our bodies. This narrative seems to fit like a glove. How can it be anything but true? There is a key distinction that requires more than a cozy narrative in order to understand. Alternative energies are science-based.

And not just that, they represent enormous scientific progress and daring technological innovation. The same cannot be said for alternative medicine, which refers to treatments that are regarded by scientists as not possessing any scientific basis. The vast majority of alternative medicine is not science-based. In fact, it typically flies in the face of basic fundamental scientific truths.

Before getting too far with this, an important distinction must be made, lest I put my foot in my mouth. I say that the vast majority of alternative medicine is not science-based, because there are rare instances in which it is. For example, marijuana has some medicinal properties. Nowhere near to the degree touted by some outlets, as it absolutely will not cure your cancer, but it may sometimes be effective in treating sleeping disorders, eating disorders, anxiety, glaucoma, and a variety of other maladies. For some time, these properties were ignored by the medical establishment, and thus marijuana qualified as alternative medicine. But there is more to the story. The fact that the adoption of marijuana as a prescription drug was delayed for such a long time, despite numerous studies demonstrating its efficacy, is related to the propaganda surrounding the classification of marijuana as a Schedule One drug, which is a classification reserved for the most addictive drugs we are aware of, such as heroin. There is no evidence to support the notion that marijuana is addictive at all, let alone to such a degree. The basis for the propaganda that had promoted it as such was purely political, and probably race-related. A deeper analysis of this example is beyond the scope of this book, so it will suffice to say that there are sometimes bureaucratic reasons why a legitimate form of medicine is omitted from the panoply of Western medicine. But this is the exception, not the rule. The overwhelming majority of alleged treatments that fall under the umbrella of alternative medicine are rejected by science-based medicine because they have no scientific basis. They are not evidence-based. In the absence of any attempt to substantiate a treatment using a scientific approach and clinical trials, one can only conclude that it is pseudoscientific. This means it is ignored not because of propaganda, but because it simply does not work, and oftentimes spectacularly so. For the sake of simplicity, when I refer to alternative medicine, assume that I am referring exclusively to these more commonly cited examples, and not the rare instances in which legitimate treatments are ignored for political reasons. We will be looking at things like homeopathy, reiki, crystal healing, and other such cultural phenomena that are universally rejected by the scientific community.

> *TL;DR—Alternative medicine is that which is not science-based.*

So if alternative medicine is not science-based, what good is it? What platform can it have to stand on? Rather predictably, and strategically, those who peddle these alternative treatments actually tend to double down on this fact. They will usually speak of "mainstream science," before alluding to a deeper tier of reality that current science has no ability to access. Whether this deeper tier involves vibrations, chakras, chi, or spirits, the nature of the claim is absurd. Science is the study of the physical world and everything in it. There is no physical object in the universe that is outside of the domain of science. If something exists in the body, we should be able to detect it, measure it, test it, and manipulate it. If such actions are impossible, there is no basis upon which to believe that such entities exist.

There are plenty of things that we can't physically see or touch that we utilize in science. Only a tiny sliver of the electromagnetic spectrum is visible to our eyes, and yet we have developed applications for X-rays, microwaves, radio waves, and all the rest. Atoms are far too tiny to see, and yet the field of chemistry flourishes. If something is real, we should be able to interact with it in some way, and the presumption that it is simply beyond the grasp of science is preposterous. It's a cop-out.

What's even more preposterous is the idea that "mainstream science" would be ignoring real phenomena, for some undisclosed reason. If convincing evidence for these alternative treatments were ever to arise, they would immediately be adopted by science and incorporated into research. Why would huge industries scoff at real treatments that could lead to real profits? Science is not immune to change. On the contrary, it thrives on change. And science is also not afraid of the realm of the bizarre and the mysterious, as narrow-minded as some make it out to be. There are countless examples in history where new discoveries left the scientific community baffled, and in awe. But if they represent real, measurable phenomena, they are pursued all the same, and they are pursued until they are understood. The entire field of physics underwent a complete transformation in the early twentieth century for precisely this reason, which we refer to as the quantum revolution. Perplexing areas of current research in astrophysics like dark matter and dark energy challenge the most brilliant minds today. Paradigm shifts are rare in science, but they do occur when necessary and substantiated.

Industries may have agendas, but science does not. Science is about what can be demonstrated, empirically and quantitatively. Conversely, until some alleged phenomenon can be studied scientifically and given a firm empirical basis, it simply is not science. So individual biases notwithstanding, the notion that the scientific community as a whole could be universally ignoring some fundamental aspect of physical reality, simply because it is stubborn, or backwards thinking, is completely absurd. Science seeks to understand the universe, first and foremost. This process is certainly guided by application, as we are most interested in pursuing the threads of scientific inquiry that will improve our quality of life. But there is a huge proportion of research effort that is simply aimed toward figuring out how things work, under the faith that applications will later present themselves. When J.J. Thomson discovered the electron in 1897, the average non-scientist would probably have replied, "Who cares?" No one alive today who appreciates the usage of their ubiquitous electronic devices would repeat this sentiment. There is no telling what applications current scientific research will later turn up. The first step is always to produce the foundational knowledge, and that is precisely what is done on a daily basis, in laboratories around the world.

> **TL;DR—Science seeks to understand the universe, first and foremost.**

Getting back to the question at hand, we were comparing alternative medicine to alternative energy. The key difference that can't be emphasized enough is that whenever alternative medicine can't be substantiated empirically, it doesn't simply bypass the pharmaceutical industry, the way alternative energy bypasses the fossil fuel industry. Alternative medicine bypasses science. Science is the domain that determines what a disease is and how to cure it, and anything that acts as an alternative to this ranges from ineffectual to downright dangerous. So we are not looking at solar panels and wind turbines here. We are looking at organizations that prey upon people with very little scientific education, who may be desperately in need, and are prone to whimsy.

Imagine that you're sick. Very sick. Potentially even deathbed sick. You're not sure what to do, and you're scared. Doctors are telling you one thing, but the internet is telling you another. Perhaps multiple doctors are giving you conflicting advice, and you're desperate for a straight answer. Even if the doctors sound reasonable and earnest, aren't they just part of the machine?

Who do they take their orders from? Hospitals want to make money, right? So why would they cure you when they could just milk you for extended treatments? There must be some solution that circumvents this whole monstrosity of a healthcare system. There has to be another way.

This is a reasonable train of thought given the emotionally charged circumstance. It is also not completely baseless. It is possible to get bad medical advice. It is possible that a particular hospital may advise useless procedures in an attempt to maximize profits. Healthcare is more expensive than it should be, in America at least. And many people have had at least one bad experience with the medical field, which is what makes it so easy for purveyors of alternatives to spin an antagonistic narrative with which to lure you away. Medical science is not perfect, and it is not immune to bureaucracy, or human error, by any stretch of the imagination. But it's what we've got. People facing serious medical problems are afraid. They're vulnerable. At times, desperate. Whether considering such a troubled individual, or one who simply empathizes with such an individual, there is no easier person to sell to, and a multi-billion dollar colossus has cropped up over the past few decades with precisely this target in mind, which we call the alt-health industry.

The primary narrative utilized by this industry is an extension of the old "natural, good; synthetic, bad" adage that we've been over several times. It uses every ounce of strength it can muster to conjure up images of man's spectacular failure to replicate the glorious, intrinsic healing powers of nature. It labels mankind's quest to understand the physical world as ignorant folly. It paints our collective scientific knowledge as uncertain, always being torn down and written anew, and most importantly, incapable of ascertaining certain deep truths about our bodies. The irony of communicating such a narrative through computers and the internet, which are made possible by our rock-solid comprehension of the subatomic world, is so thick that you could cut it with a knife.

With all of the incredible innovations of the past century, why might people still be drawn to the notion that nature cures best? The answer is simple. For most of recorded history, nature was all we had. And we tried literally everything. Every type of herb, leaf, root, bark, and fungus we could find. You've got a rash on your skin? Rub this on there. You're having stomach pains? Drink this and eat that. Thousands of years of blind trial and error. And what would happen? Sometimes people would get sicker, or even die. Usually not much of anything would happen. However, on rare occasion, it would be discovered that something found in nature had medicinal properties. Take Cinchona trees, for example. Certain species of this genus contain bark that was used to treat malaria, upon the discovery of this ability in the seventeenth century. In absence of any mechanistic knowledge, one could be tempted to presume that the medicinal properties are held by the tree itself. This tree could be labeled as special, or sacred. In actuality, we now know this bark contains a compound called quinine, which belongs to a class of molecules called alkaloids. Quinine is the active ingredient of the bark that does the relevant chemistry, killing the parasite that causes the disease. We can isolate this compound from the bark, administer it to a patient, and observe that it does the job all on its own, something that was first accomplished around 1820. But furthermore, quinine can also be synthesized in a lab, which was achieved in 1944. Each step of the way, the phenomenon is further demystified. From a sacred tree, to a sacred component, to a mere chemical process.

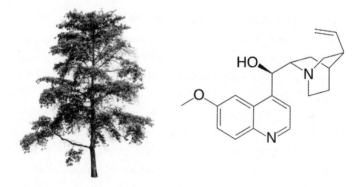

The wild Cinchona tree (*Anthocephalus chinensis*) and the component of its bark that possesses anti-malarial properties, quinine.

*TL;DR—Plants with medicinal properties are not magic, they contain an active ingredient, some molecule that does specific chemistry which can be understood.*

This is actually a pretty good way to talk about the origin of the pharmaceutical industry. Some challenges for early chemists involved isolating and characterizing these compounds, which were found in living organisms and had medicinal value. We had to figure out which molecules did which things. Once this was achieved, the next challenge involved finding reasonable synthetic pathways, such that these compounds could be synthesized in a lab from inexpensive and commonplace starting materials, rather than continuously hunting for natural sources, extracting, and purifying. This practice lasted quite some time, but in the twentieth century, chemistry evolved far beyond the limited practice of mimicking nature. Nature has stumbled upon some interesting compounds by blind chance, no doubt. But with actual sentience guiding the process, the game had changed. Chemists realized they could do better than nature and began to create completely new compounds in a target-specific manner. Humanity began to understand how the body works. We began to understand what is happening on the molecular level when something goes wrong in the body. And we began to have the courage to come up with our own solutions, sometimes involving completely novel compounds that have never existed in nature, whose properties we were able to predict and verify, thanks to a modern understanding of chemistry, physics, and computer science.

The elegance of this impressive achievement tends to be lost on the masses. Our ability to manipulate matter on the molecular level is no match for the romance of herbalism. This is the practice of stubbornly clinging to ancient inclinations toward the sacredness of the plant. It somehow seems so personal, so comforting, that we could be healed by a plant. Perhaps it's because they are also beautiful. Perhaps it stems from a belief in the divine, and its superiority over us mere mortals. And of course, it also stems from an ounce of truth, because plants have historically been used to heal. But when the plant itself is placed on a pedestal, over the actual chemistry that plant is known to promote, there is only ignorance to blame for the rejection of centuries of scientific progress. It is yet another echo of vitalism, a baseless insistence that only the plant can heal. The inability to accept that in most cases, the extract of the singular compound with medicinal properties will function identically, as will a laboratory-synthesized version of that compound.

I do not mean to discount modern herbalism that is based on economic reasons. If growing a particular plant to harvest the medicinal benefits is cheaper than purchasing drugs, particularly in certain poverty-stricken areas of the world, then this remains a fair course of action. Again, the plant contains the drug, and we can make the drug. It's just some molecule, which is no better or worse whether encountered in nature or in a pill. Choosing one route over the other to be economical is no problem. It is only the stubborn belief that one is valid and the other is not that should be in the crosshairs.

But this is all pretty small potatoes. You've read the chapters that led us to this point, and presumably you're on board. A molecule is a molecule, so on and so forth. We get it. Now let's tackle a trickier beast, the pharmaceutical industry. In the eighteenth century, drug-making duties fell upon a class of professionals that were referred to as pharmacists, chemists, and apothecaries. These people prepared remedies in their labs following some general recipes, formulating them in elixirs, ointments, and powders. These were served directly to the customer, often bypassing doctors, who could advise on therapies but were too expensive for the masses to afford. Then in the nineteenth century, a new industry began to create new therapies and manufacture more effective drugs on an immense scale. The end result of this shift is that pharmacists now simply receive a doctor's prescription and deliver the listed pre-packaged medicine to the patient for a tiny profit. The story of this revolution is the story of the pharmaceutical industry. So what is this industry all about today? We know it is widely criticized, but for what exactly? Are these attacks legitimate? The short answer is both yes and no. But unlike alternative health claims that have nothing to do with science, current medicinal practices can be discussed and optimized, so let's try to understand how this industry operates.

The concept of drug discovery and synthesis is exceptionally esoteric to most people. Even with the information we've gone over thus far regarding chemistry and biochemistry, we are barely equipped to scratch the surface on this overwhelmingly complex practice. But let's give it a shot anyway, in our typically reductive fashion. Consider some disease that humanity is interested in curing. As chemists invested in developing a drug to this end, what do we do?

*Step 1: What is the cause of the disease?*

As we now know, all disease has a molecular basis. Let's say that biological studies have identified a potential target for the cause of some disease, which is a particular protein. This protein is misbehaving in one way or another, and this is the fundamental cause of the symptoms of the disease.

*Step 2: What do we do about it?*

Now it's time to select a strategy. We may elect to come up with a small molecule that will interact with our target protein and modify its behavior. What will this molecule be? Let's take a million or so molecules and screen them for binding affinity. We get a few dozen hits. Then we recognize some structural features that seem to be common to all of them, and we begin to refine the structure, aided by X-ray crystallography and computer simulations. Eventually, we settle on a structure for the potential drug.

*Step 3: Synthesize the drug.*

We have a structure. Now we figure out how to synthesize it from cheap, readily available starting materials. How shall we do it? What kind of pathway shall we employ? The possibilities are infinite, and comprehensible only to the trained synthetic organic chemist. But it has to be as cheap and efficient as possible, with the fewest steps possible.

*Step 4: Scale it up.*

Congratulations, you've got a pathway. But guess what, that tenth of a gram in a vial isn't going to be enough to help anyone. We need literal tons of this thing. In process development, the synthesis is scaled up using enormous industrial machinery, and there are many aspects of this that require specialized techniques. One batch may cost millions of dollars, so there is very little room for error here. The process must be flawless, achieving highly reproducible quality and purity, which is exceptionally difficult and requires years of effort.

*Step 5: Start with the clinical trials.*

It may be surprising to some, but drugs are not unleashed on the market willy-nilly. On the contrary, the legalities regarding clinical trials are extremely stringent, and for good reason. While we can be very confident regarding the ramifications of introducing a novel substance into the body by performing experiments on isolated proteins or entire cells, we are never completely certain what will happen until we perform tests on living systems. Animals first, then humans, with strict regulations every step of the way.

*Step 6: Patent, market, and sell.*

If everything went smoothly, congratulations! You've made a drug. A patent has certainly already been filed, because you've just invested an obscene amount of money, and you are legally allowed a temporary margin of exclusivity in the sale of this product in order to recoup expenses and make a profit, just as one would expect in any industry. Name the drug something catchy, and market it however you wish, within the bounds of regulation.

In the tiniest nutshell imaginable, that's pharma. So what's with all the hate? Let's investigate. First, criticisms are voiced regarding the profit margins on pharmaceutical drugs. This one is easy to combat. There is no industry on earth that takes on greater risk than pharma. At any of the steps outlined above, the whole process can suddenly become a bust. Perhaps excessive screening affords no convincing hits. Perhaps an efficient synthesis is elusive. Perhaps an issue arises when scaling up. Perhaps during clinical trials, it becomes apparent that the drug is binding to its target flawlessly, and yet symptoms persist, which means the whole thing was a wild goose chase for the wrong target. Perhaps unforeseen side effects crop up that make the drug toxic and therefore unmarketable. Perhaps on this timeline, which can sometimes be a decade long, another company is only six months ahead and beats you to the punch with a drug that is just as good as yours. If any of these events occur, a company will stand to lose hundreds of millions of dollars and have absolutely nothing to show for it. And these occurrences are not rare. Only one out of ten drugs makes it to market. That means that on average, a pharmaceutical company has to spend over two billion dollars to produce a profitable drug. That's a lot of money to have to make back. And that's the explanation for the profit margins. Of course there are those who abuse this, but the average net profit margin of 20 to 30 percent on a drug is completely reasonable in the context of what we have discussed here, particularly considering that once introduced to the market, a drug can only maintain exclusivity for ten to twelve years at most. Manufacturers of other products with traditionally much narrower profit margins do not deal with such immense risk, because even if the product is unpopular, it still has some intrinsic value, thus offering some ability to partially recoup on investment. If a chair isn't selling for the luxury price you had hoped, despite everything you sunk into advertising, you can always sell it for half as much, in a worst-case scenario. With drugs, if it's a bust, you are left empty-handed.

Then there is the criticism of patents. Let's reiterate how much money is required to bring a drug to market. The risk of failure is high. If that risk were compounded tenfold by the prospect of having the fruits of your research snatched from under you, such that a rival company could piggyback and beat you to market with your own drug, nobody would make drugs. Patents are an absolute necessity in industry, and particularly in drug development. Depending on the precise timeline, companies typically have around a decade of exclusivity on a drug, within which they can make substantial profits. After this period, anyone who wants to make a generic version of that drug is free to do so and undercut the original manufacturer however they can. This is quite feasible because they don't have to do any research to demonstrate the safety and efficacy of the drug, as it has already been done. Anyone else can now optimize the process, make the exact same drug, and simply call it something else, since names can be trademarked, unlike molecular structures. If successful, they can then sell the generic yet identical version at a lower price. This ensures that greed is no longer a factor, because the drug is no longer monopolized and suddenly enters a competitive marketplace. If someone else can charge less and still make a profit, they absolutely will do so. This also destroys any paranoid notion of the pharmaceutical industry "hiding cures" to diseases. This industry is not a singular entity. It is not one old man sitting behind a desk in a room, plotting and scheming. It is a dynamic and competitive landscape, and one company's missed opportunity is another one's claim to fame.

Hopefully we are starting to get a better sense of how the pharmaceutical industry operates, and why many attacks are misled. Just to keep things even-keeled, there are some legitimate criticisms. For one thing, companies exclusively choose targets that they believe will be profitable. This means focusing on diseases that are suffered by many people rather than few. The result is that certain diseases get largely overlooked or ignored, because there isn't enough financial incentive to pursue a cure. This is unfortunate, because in a perfect world, we would work on curing every disease. This gives rise to what are referred to as "orphan drugs," and one solution has been for the government to offer financial incentive in exchange for working on drugs that would generate less profit due to the lower demand. This does indeed produce a flurry of activity, particularly amongst small biotech companies.

Of course, we could expand on this line of thinking and acknowledge that the alternative to the pharmaceutical industry, with its high risks, exclusivity, and generous profit margins, would be entirely state-sponsored research. This would mean that the cost and risk associated with pharma research, around a hundred billion dollars in 2019, would fall entirely to the taxpayer. This is how drug development operated in Russia until the end of communism, and the number of life-saving drugs that were produced is not impressive. This is just one of the many reasons why human civilization is destined to forever walk the tightrope between governmental and corporate dominance in human affairs. We need both sectors to exist, and neither should have total control over the other.

Beyond this, there are admittedly some ethical transgressions. Incentives have been illegally provided to doctors in exchange for favoring prescription of a particular drug, or for championing off-label use of a company's product line, so as to increase the range of indications for which the drug may be used. Lobbyists influence legislation that favors industry. And there are undoubtedly plenty of instances in which corporate greed rears its ugly head. The pharmaceutical market is now worth about a trillion dollars per year. As with every economic sector that commands this amount of money, whether looking at banks, transportation, weapons, or medicine, corruption will exist. But the answer is not to abandon progress. There must be a measured effort to remain vigilant and empower regulatory agencies which can act as watchdogs in their respective sectors. Just as with every other sector, pharma companies have had to deal with huge fines and closures when they have crossed the line, and we must remain steadfast in this practice, while simultaneously acknowledging that every aspect of this discussion is entirely outside of the domain of science.

> *TL;DR—Criticisms of the pharmaceutical industry are largely, but not entirely misled.*

I don't mean to digress into political commentary, nor do I intend to make your head spin with a deep dive into the specifics of this industry. But a degree of explanation is necessary. Understanding how legitimate medicine works will allow us to dissect certain forms of alternative medicine and see how laughably they fall apart with even the tiniest bit of scrutiny. The biochemical studies, the clinical trials, none of the rigidity and empiricism of drug development is present in such practices. So let's set up some cans and knock them down.

First up is the mascot of alternative medicine, homeopathy. Many have heard of it, and few know precisely what it claims. In short, homeopathy operates under a framework that can be summarized as "like cures like." If someone is ill, whatever substance caused the problem, that's also going to be the solution. A patient's physical, mental, and emotional states will be assessed, and the substance that will act as treatment will be chosen. Then once chosen, the substance will be diluted. And then shaken. And then diluted again. A lot. By a factor of around a million trillion trillion. The results are then consumed, and that's all there is to it.

So if that's truly the extent of the process, where is the efficacy? Well the more dilute, the more potent, according to homeopathic practitioners. But how is that possible? Well, it isn't, unless we want to declare the entire edifice of chemistry as null and void. There is no logical framework in which any of this makes sense. If some compound does something in the body, the more of it there is, the more it will do that thing. In actuality, the reason homeopaths dilute so much is so that the treatment won't harm you or produce any side effects, because there is truly nothing in there, not even a single molecule of the original substance. They also issue disclaimers not to use this kind of treatment for life-threatening illnesses. This is because it doesn't actually do anything, and you'll die. Even charlatans cover their bases. The claim that shaking a vial can make its contents more potent is ridiculous. The notion that if a compound causes harm to the body, then the very same compound can then reverse that harm, has no merit whatsoever. Homeopathy is immediately recognized as patently absurd by anyone with a high school level knowledge of chemistry and biology. In fact, the practice only gained popularity upon its inception in the eighteenth century because a treatment that does absolutely nothing was far superior to certain alternatives of the time, like bloodletting, which specifically harmed the patient. It predates knowledge of pathogens like bacteria and viruses. It predates the entire field of modern chemistry. It's nothing more than a relic from a pre-scientific approach to medicine.

> ### *TL;DR—Homeopathy is placebo.*

As we can clearly see, this is an instance where even just the minimal amount of scientific information that has been presented in this book, which bestows the reader with only the most rudimentary comprehension of biochemistry, is more than enough to conclude with complete certainty that homeopathy is hogwash. There is absolutely no room to claim that science simply doesn't understand it, or that some deeper aspect of reality eludes the scientific community. The claims made by homeopathy fly in the face of the most fundamental and indisputable facts of the chemical world. There is no example I can think of in a medical context that better defines pseudoscience than homeopathy. In later chapters, once we define energy and other concepts of physics, we will be able to apply the same kind of analysis to things like reiki and crystal healing that will elucidate the sheer absurdity of these practices in precisely the same manner.

To go beyond simply referencing the science that homeopathy contradicts, it should be stressed that every single clinical study of homeopathy has failed to detect any effect whatsoever beyond placebo, meaning a mild and fleeting mitigation of symptoms that are modulated by the brain, which is brought on psychologically due to expectation. Anyone who claims that the treatment has worked on them is undoubtedly citing a placebo effect, which is not to be severely underestimated, particularly in cases where a disease is psychosomatic to begin with. Psychological effects can be quite significant. But they are not causally linked to the contents of homeopathic treatments. They can arrive just as well from someone waving a magic wand and shouting "abracadabra," if one is prone to believing that should work. There is a reason we utilize placebo as a reference point in clinical trials. A treatment must perform better than placebo in order to be considered effective, and homeopathy does not do this. For this reason, it is perplexing that homeopaths have at times teamed up with credible scientists to advance hypotheses such as the notion that efficacy stems from the "memory" of water, asserting that water can remember the shape of what was in it, and this shape somehow has a physiological effect. This is preposterous, and not supported in the slightest by science. What is even more perplexing is that a handful of qualified scientists end up working in areas like this, but let's remember that scientists are human beings. They have ambitions, desires, and delusions just like the rest of us, and even the most respected Nobel prize winners can sometimes go off the deep end into the pool of pseudoscientific nonsense.

To tie all of this in with a previous point, the notion that pharmaceutical drugs treat the symptoms while alternative medicines treat the cause, can now be appreciated for the complete reversal of reality that it is. Homeopathic treatments offer a placebo effect and nothing more, and they are indeed often administered with the exact same sugar pellet used for the control group in a clinical trial. They do nothing whatsoever to address your health on a fundamental level, and brief psychosomatic alleviation of mild symptoms is all they have to offer. One could argue that this is at least something, and with reasonable basis, as any pain relief is legitimate relief even if by completely internal means, but this is certainly not what is advertised. An appeal to romance, an evocation of the old and the ancient, these are marketing tactics that should not be equated with truth or wisdom. Remedies that predate modern science also predate our understanding of illness and the human body. With medicine, unlike wine, older is not better. Claims to the contrary should be seen for what they are, an attempt to take your money and give you nothing in return, or at the very least, profit-driven compliance with client demand, which is immoral for any medical professional.

> **TL;DR—Ancient medicinal practices do not equal wisdom. It is typically the contrary.**

The alarming thing is that public interest in alternative medicine is growing so strong that we are seeing sectors of education bend in response. There are institutions of higher learning that offer formal programs of study in homeopathy. That should terrify you. This is not some dark corner of the internet. That such institutions would recognize an opportunity for profit, and offer programs to cash in on the demand, is an embarrassment. It is the admission that profits turn the gears in every sector of our society, and it will require a massive shift in the social consciousness to rectify this. Let's make sure that such a shift is coming.

# CHAPTER 7

# **The Body as Machine**

Up until this point, we've focused entirely on molecules. That's because our bodies are nothing but molecules. But the whole is always greater than the sum of its parts, and in order to understand how the human body functions, we are going to have to climb up another rung on the ladder of complexity. Just the way that we discussed some basic principles of chemistry and biochemistry, so that we were equipped to comprehend relevant concepts in those domains, we will have to put a dent into basic biology principles in order to have any hope of understanding human physiology and general health. So from small molecules, to big molecules, we increase the scale once more to examine **cells**. These are the functional units of life. What are cells made of? What do they do? How do they differ from one another, and how do they become that way? How do they work together to produce and maintain that person you see in the mirror every morning? It goes without saying that what we are about to discuss in the next few pages will be far from comprehensive. But going over the bare minimum regarding cells and cellular function will allow us to talk with some sophistication about things like cancer, which is undoubtedly of universal concern, so let's get started.

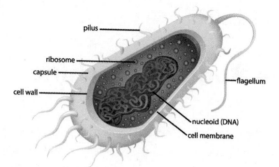

*pilus*
*ribosome*
*capsule*
*cell wall*
*flagellum*
*nucleoid (DNA)*
*cell membrane*

The structure of a typical bacterium, which is an example of a prokaryotic cell.

Cells are little bubbles of life. The smallest things that are considered alive, on earth anyway, are unicellular organisms, which means they consist of just one cell. Bacteria are examples of such organisms, and bacterial cells are very simple cells called prokaryotic cells. There isn't much in there, just a circular chromosome containing all of the bacterial DNA suspended in the middle, some enzymes and ribosomes for gene expression, lots of small molecules floating around in the cytoplasm that fills up the cell, and some other molecules to serve as the boundary between the organism and its environment, which we will discuss in a moment.

Some unicellular organisms are not prokaryotic, but rather eukaryotic, meaning that they are made of one eukaryotic cell, which is much more complicated. These kinds of cells have lots of little structures inside called organelles, which are to a cell what our organs are to our bodies. Each organelle has a particular function. The nucleus, as we know, holds the genetic information. The ribosomes build the proteins, as we also learned. The endoplasmic reticulum folds up the proteins. The Golgi apparatus tags them and sends them to where they need to go. The mitochondria make the energy, which we will get to later. Every organelle has a role, which together make eukaryotic cells larger and more complex than prokaryotic cells.

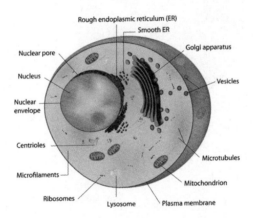

The structure of an animal cell, which is a type of eukaryotic cell.

All multicellular life is made from eukaryotic cells. This includes all plants and animals. Let's look specifically at animal cells, since we are animals. In these, all of the organelles, and the entire cell as a whole, utilize molecules called phospholipids to act as membranes. Remember when we mentioned the four classes of biomolecules? Those were proteins, nucleic acids, lipids, and carbohydrates. Well, phospholipids qualify as lipids, naturally, and they are crucial to the existence of cells, because you can't have a cell without some membrane to separate what's inside the cell from what's outside the cell.

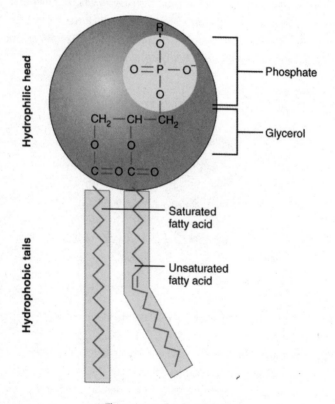

The structure of a phospholipid.

Let's quickly describe these phospholipids. Their key characteristic is that they are amphiphilic. What this means is that one side of the molecule interacts very well with water. This is because of a negatively charged phosphate group, like the kind we saw in nucleic acids, and this negative charge can make electrostatic interactions with the part of a water molecule that bears a partial positive charge. The other side of the molecule does not interact well with water, because it is greasy, or nonpolar, made simply of hydrocarbon, having no partial charges with which to make such interactions.

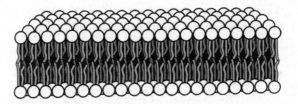

Phospholipids, when in water, self-assemble into bilayers.

Because one end likes to be near water and the other end does not, this allows phospholipids to self-assemble into something called a bilayer. This is comprised of two layers of phospholipids, and all the phospholipids within each layer point the same way. Specifically, the phosphate heads point outwards and the hydrocarbon tails point inwards. The reason this occurs is because this is the configuration that maximizes the interactions with water that these molecules can make, as the nonpolar tails do not interfere with such interactions if they are hidden away from the water. So we have water outside of the cell interacting with the outer layer, and water inside of the cell interacting with the inner layer, and a nonpolar region in between that consists of the tails from both layers of phospholipids. All the cell organelles have membranes like this as well. It is worth stressing that when phospholipids are placed in water, they will spontaneously form these structures. No energy input or enzymatic activity is required. It is the hydrophobic effect that causes the hydrocarbon tails to hide away from water, which is simply a ramification of a system of molecules striving to reach the lowest energy possible.

> *TL;DR—Phospholipids spontaneously organize to form the boundary of a cell.*

Within the cell membrane, also called the plasma membrane, there are many surface proteins. Like little rafts on the ocean, these proteins float around in the membrane, which is rather fluid. The proteins serve a variety of functions related to cell communication and other such things, which we will touch on a bit later. And we say that the membrane is semipermeable, meaning that some things can get through, while other things can't, depending on how polar they are. Small, nonpolar molecules can squeeze in between the phospholipids and sneak in or out of the cell. Large molecules as well as small polar molecules can't, because they either don't fit, or they can't traverse the nonpolar region. If they are needed inside the cell, the cell can bring them in through various channel proteins, which are like little pores that allow molecules to pass through, using different mechanisms that may or may not require energy to occur.

The plasma membrane contains many membrane proteins with different functions.

Some of the surface proteins floating around in the membrane are called receptor proteins. These receive chemical signals from other cells when a specific molecule, called a ligand, binds in lock-and-key fashion in the active site of the receptor, just like an enzyme and its substrate. There are also receptors inside the cell, or even within the nucleus of the cell, which have ligands that are able to get through the plasma membrane in order to reach their target.

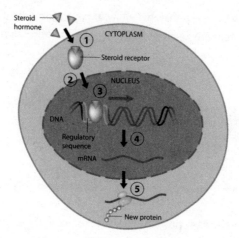

Scheme illustrating the activity of a transcription factor,
which is a type of receptor.

With certain receptors, binding occurs with the ligand that acts
as a signal, which is typically a molecule called a hormone. As a
result, a conformational change occurs in the protein that allows
it to bind to a particular section of DNA next to a gene, which
in turn allows a polymerase enzyme to transcribe that gene.
These receptors are called transcription factors, and this is one
of several methods that are utilized to regulate gene expression,
which is a phrase that implies a way of controlling which genes
are transcribed and when. This is already getting immensely
complicated, so we won't go any further on that front. We now
have a reasonable picture of a cell, so what's next?

As we said, humans are multicellular animals. An adult person contains somewhere in the ballpark of a hundred trillion cells. That's not quite as many as the trillion trillion molecules in your glass of water, but it's an enormous number nonetheless. So how does this work? These cells certainly are not all the same, or we would just be one huge amorphous blob. How is it that we have limbs and organs? How is it that we can run, and jump, and think? There are, of course, many different kinds of cells. Every cell has precisely the same DNA, but the specific genes that are expressed in any given cell can be different. During the development of an embryo, certain things called cytoplasmic determinants are distributed in such a way so as to cause cells in particular regions to differentiate in a specific manner. That is, certain genes are expressed that govern the characteristics of each cell. This includes the shape, elasticity, and function, depending on which proteins are produced by that cell and to what degree.

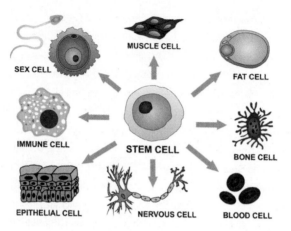

Cells differentiate to form different cell types.

*TL;DR—Cellular function is determined by gene expression.*

These different kinds of cells come together to form **tissues**, which are collections of cells that are similar in structure and perform a common or related function. The four types are epithelial tissue, connective tissue, muscle tissue, and nervous tissue. Epithelial tissue makes up the outer layer of your skin and lines your bodily cavities. Connective tissue provides a protective, structural framework for other tissues, including cartilage and bone. Muscle tissue makes up all your muscles, which obviously include the ones that allow you to move your body around at will, but they also regulate involuntary motion like your heartbeat, as well as the constriction and dilation of various vessels and organs in order to push substances around the body. And nervous tissue consists of the neurons all around your body that allow for signals to travel to and from the brain, as well as assorted other cells that maintain them.

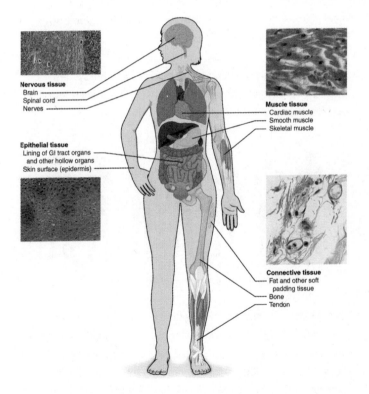

**Nervous tissue**
Brain
Spinal cord
Nerves

**Muscle tissue**
Cardiac muscle
Smooth muscle
Skeletal muscle

**Epithelial tissue**
Lining of GI tract organs and other hollow organs
Skin surface (epidermis)

**Connective tissue**
Fat and other soft padding tissue
Bone
Tendon

The four types of tissue in the human body.

These four types of tissues come together in various ways to produce all of our organs and organ systems. These are regions of the body that are specialized for a particular function. There are the lungs and the respiratory system, bringing oxygen from the atmosphere into the body. There is the heart and the circulatory system, pumping blood throughout the body so that oxygen can get to every last cell, and so that every cell can dump out its waste for removal from the body. There is the nervous system, producing all of our sensations and their subsequent perceptions in the brain. There is the skeletal system, the muscular system, the digestive system, the endocrine system, the immune system. There are a lot of systems. Humans are complicated.

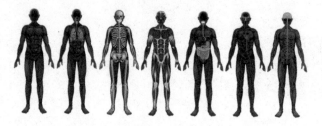

Some of the organ systems of the human body.

Let's be clear. If we were to expand on everything we just mentioned to any considerable degree, we could learn a lot more about the body, but that just isn't our purpose here. This is not a textbook; it is social commentary interspersed with brief instructional passages. If what we've gone over together so far has inspired you to seek a deeper understanding of all of this stuff, from tiny molecules to human anatomy, I encourage you to continue on your path of learning as soon as you finish this book. But from this point on, we will merely reference certain aspects of our anatomy as necessary. With some basic understanding of structure, let's move on to function. What is it that cells do?

One thing that cells have to do is generate energy. Every time you move around, or breathe, or even simply think, your body is using energy. Where does this energy come from? Well, it all starts at the sun. The sun shines on plants, which through a process called photosynthesis, use light and water and carbon dioxide to build glucose, which belongs to a type of molecule called sugars.

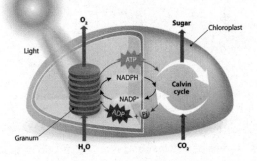

Plants perform photosynthesis, which produces glucose, a sugar.

Then we eat the plant, or we eat an animal that ate the plant, and molecules such as glucose enter the digestive system, which are then absorbed into the bloodstream and shuttled around to all the cells in the body. From here, a process called aerobic respiration takes place. This consists of three steps. The first is called glycolysis, which occurs in the cytoplasm of the cell. Then there is the citric acid cycle, or Krebs cycle, and the electron transport chain coupled with oxidative phosphorylation, all of which occur within mitochondria, one of the cell organelles.

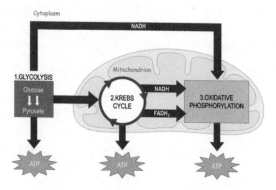

Aerobic respiration is the primary method of energy production in animals.

This three-part process is so overwhelmingly complicated, involving dozens of enzymes and chemical reactions, that we won't even attempt to elucidate them here. But the end result is the conversion of many molecules of adenosine diphosphate, or ADP, into adenosine triphosphate, or ATP. These look familiar because they are like the adenine nucleotides we find in nucleic acids, just with either two or three phosphate groups instead of one. And it is ATP that is the cellular currency of energy.

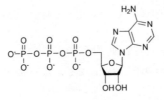

Adenosine triphosphate, or ATP. Removing one of the three phosphate groups would produce adenosine diphosphate, or ADP.

We will define energy in all its forms with some rigidity a bit later. For now let's simply mention that when there are three phosphate groups present, in the case of ATP, the repulsion between the negative charges in those phosphate groups raises the potential energy of the molecule, kind of like a compressed spring. And when the outermost phosphate group is transferred to some other molecule, during what is called a phosphorylation reaction, some of that potential energy is released. This is analogous to releasing that compressed spring, allowing it to expand, where the energy from the "expansion" can drive some other cellular process. Virtually every cellular process that requires energy is driven by the conversion of ATP into ADP, and then the immaculately complex yet completely automatic series of reactions that comprise aerobic respiration will phosphorylate ADP to get back ATP again. This is why we eat food, and it is also why we breathe oxygen, because the third step in aerobic respiration is impossible without oxygen, as the combustion of sugars is an oxidative process. This necessity explains why we die so quickly without oxygen, as without a means of energy production, we can't do any of the moving or breathing or thinking that was mentioned a moment ago.

*TL;DR—ATP is the molecule that powers most cellular processes.*

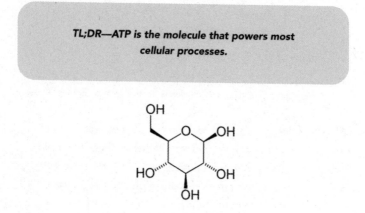

Glucose, a monosaccharide.

Once again, glucose is a sugar, and sugars are a type of carbohydrate. This is the fourth and final type of biomolecule we need to understand. Glucose is specifically a monosaccharide, and these can polymerize to form polysaccharides, which are another type of biopolymer. One form of glucose called beta-glucose can polymerize to form cellulose, which gives structure to plants, though we are unable to digest it because we lack an enzyme that cows possess. Another form of glucose called alpha-glucose can polymerize to form starch, which we are able to eat and break down using enzymes to produce individual glucose units for cellular respiration. We also polymerize glucose ourselves to store it as glycogen, for later usage if food sources become scarce. In addition, we can see shorter saccharide chains used in various other ways within a cell. Sometimes they are attached to certain proteins in the cell membrane, and this is one way that cells can communicate with one another, through recognition of these groups by other molecules.

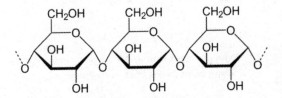

Glucose can polymerize to form cellulose, starch, or glycogen.
Starch is shown here.

There is so much else we could discuss. There are small molecules called hormones that are released by glands and travel through the bloodstream to reach specific target cells, inducing conformational changes in receptor proteins that elicit some physiological response from the cell. We could talk about the way red blood cells utilize a protein called hemoglobin to carry oxygen from the lungs to every bodily extremity. We could talk about the way neurons utilize chemical potentials to transmit electrical signals from your sensory receptors, where we sense our surroundings, to the brain, where that sensation is processed into genuine perception, which then sends instructions back out to other areas of the body in response to the perception. We could talk about the mechanism of muscle contraction, which is activated when the brain decides it wants to move a muscle. It cannot be overstated how incredibly complex each of the above processes is, and how thoroughly satisfying it is to develop comprehension of them, until human physiology no longer seems like magic, but rather can be seen for the grand symphony of chemical reactions that it is. We are organic machines, in the most literal way possible.

> *TL;DR—Even the most complex biological processes are nothing but a series of chemical reactions.*

In absence of more thorough comprehension, because we don't have another ten thousand pages to fill, I will simply stress one thing. The human body, as we have discussed, is not perfect. There is no "natural state," or chemical-free purity that it longs for. The body has faults, it has weaknesses, it is imperfect, both in structure and in function. It is the product of several billion years of evolution by natural selection, the perpetuation of solutions to environmental problems that were stumbled upon by blind chance, and it should not be likened to the supernatural. It is quite romantic to think of the human body as perfect, and we are often marketed this notion, in the hopes that we will purchase products in an effort to get back to this elusive state, which we could successfully perpetuate if only we could rid ourselves of these pesky toxins and harmful chemicals.

We, too, can look and feel like this athletic and fashionable couple hiking up a mountain. We, too, can commune with our inner divinity and transcend the disgusting, shameful, decaying meat sacks we see staring back at us in the mirror. And all for just four easy payments of $19.99. With a slightly enhanced understanding of the human body, from the cellular level and upwards, we are ready to dive even deeper into the realm of the internet hoax and see just what we turn up.

# CHAPTER 8

# Recognizing Science-Based Medicine

We talked about small molecules. We talked about biomolecules. And now, after a brief introduction to cells and human physiology, we are in an even better position to discuss health and medicine. What happens when certain cellular functions are disrupted? What strategies can we employ to remedy such situations? Science has plenty to say about this, as does the alt-health industry, so let's dive a little deeper and see what we can elucidate.

Why not start with the granddaddy of them all, cancer. What is cancer, exactly? To answer this, we need to know a little about DNA replication and cell division. As we know, we are made of trillions of cells. But we all started out as just one cell. When your dad's sperm fertilized your mom's egg, it produced a zygote. This was the first cell that ever existed with your precise genetic material, two distinct sets of twenty-three chromosomes, one from each parent, for a total of forty-six. In order to build a human, more cells had to come about, which means this cell had to multiply, so to speak. This happens via a process called **mitosis**, whereby one cell becomes two. For this to be achieved, copies are made of everything in the cell. Certain steps then occur which allow the copies to segregate to either side of the cell, and then the plasma membrane pinches off down the middle to create two distinct daughter cells, each with the same genetic material as the original. Every cell needs a copy of all the DNA, and this is made possible by DNA replication.

DNA replication is a process involving many enzymes.

The process itself is carried out by a host of enzymes, which are different from the ones that transcribe the genes. These enzymes work together to unzip the double helix of every chromosome, pry the strands apart, and synthesize complements of each strand. The complement of the first original strand will be identical to the second original strand, and the complement of the second original strand will be identical to the first original strand, so when this process is complete, there will be two copies of every chromosome. When we see pictures of chromosomes in the familiar shape, looking like arms and legs attached at a junction, this is actually a chromosome that has completed DNA replication to produce two sister chromatids, in which the arm and leg to the left are completely identical to the arm and leg to the right. This is different from homologous chromosomes, which refers to the two versions of a particular chromosome, one from the father and one from the mother, which contain the same genes, but whose sequences are not identical.

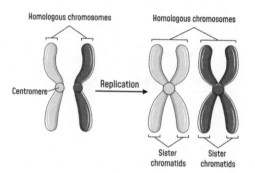

Chromosomes undergo replication, resulting in identical sister chromatids.

Then during mitosis, the chromosomes line up in the middle of the cell, and certain proteins are responsible for separating the sister chromatids and pulling them to either side of the cell, so that when cell division is complete, each daughter cell has one unduplicated version of each chromosome. Later, replication will take place again, so that cell division can take place again when necessary.

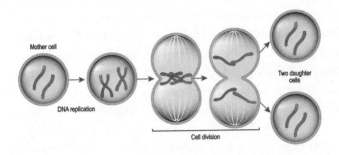

Mitosis is a form of cell division whereby one parent cell produces two identical daughter cells.

Naturally, during embryonic development, cells are dividing constantly, because of the rapid growth that occurs during this time, in order to go from a single-celled zygote to a fully formed fetus toward the end of pregnancy. In childhood, again cell division is quite frequent in most parts of the body, as we need more cells in order to grow. But cells must not always be constantly dividing. There is a point where bodily growth halts, and beyond this, there are certain types of cells, like neurons, that essentially never divide. So there is an exceptionally complex system of signaling events that regulates when specific cells should be dividing. This is called the cell cycle. The presence or absence of small signaling molecules will determine whether a cell enters a non-dividing state, or performs DNA replication in preparation for cell division, and the presence or absence of these signals corresponds with things like certain periods of growth, or entry into adolescence. However, as one can imagine, there are very many proteins that are associated with all of this. Receptors are involved in receiving the signals. Enzymes are involved in DNA replication. Enzymes are involved in cell division. Enzymes are involved in regulating the expression of the genes that code for all of the enzymes that regulate cell division. It's complicated. And if mutations occur in any of the genes that code for any of these proteins, the mutant protein may have its function altered in a way that is disastrous for the cell. Usually, when something goes irreparably wrong within a cell, it can initiate apoptosis, which is sort of like a self-destruct mechanism. The cell commits hara-kiri for the benefit of the organism. But there are lots of proteins that are involved in this process and its regulation as well, and mutations can occur in the genes that code for these apoptotic proteins too. This could render cells incapable of carrying out apoptosis when they ought to, which is problematic.

Whenever significant mutation is sustained, it is possible that related protein function may be altered such that cell division occurs when it isn't supposed to. Rather than following instructions to remain static, cells continue to divide until a tumor is produced, which is simply a mass of cells that is caused by abnormal growth and serves no physiological purpose. That's it. It's just extra cells that aren't supposed to be there. If this mass of cells does not intrude into other tissues or disrupt normal bodily function, we would call the tumor benign. This could be removed by surgery and is unlikely to cause problems. But if the tumor invades other tissues in harmful ways, we would say the tumor is malignant, and this is called cancer. Any type of cell can sustain mutations, so any type of cell can become cancerous. And if some of these cancer cells were to metastasize, which means they spread around the body, usually by slipping into the bloodstream and attaching to other tissues to continue to grow, this becomes much more difficult to treat, as current technology has very little ability to discern between a cancer cell and a normal cell, particularly if the mutant proteins are not on the surface of the cell, such that they can't be explicitly recognized.

> **TL;DR—Cancer is just cells dividing when they aren't supposed to.**

What we must understand is that mutations are inevitable. Apart from our ability to inherit certain genetic predispositions toward cancer, we also sustain mutations on a daily basis. If exposed to mutagens, the rate of mutation is much faster, in the case of spending too much time in the sun, or in extreme scenarios when exposed to a source of high-energy radiation. But our bodies emit radiation naturally, in ways that we will elucidate a bit later. And the enzymes that perform DNA replication make spontaneous errors as well. Sometimes these mutations and mistakes are fixed, but sometimes they aren't, which means it is absolutely inevitable that errors in the genome will slowly accumulate. It is only a matter of time before such errors crop up within genes that code for some of the aforementioned proteins, and cancer occurs. So cancer is the great limiter of biological life. On a long enough timeline, any living organism would eventually get cancer, as it is impossible for an organism's genome to remain pristine indefinitely. Nature is flawed. And after trillions and trillions of replications, no genome can possibly be precisely the same as what was present in the single-celled zygote that started the whole show.

All of this is crucial to grasp for two reasons. First, we have to know what cancer is on the cellular and molecular level in order to discuss a science-based approach to treatment. It is tempting to view cancer as a bunch of cells with frowny faces on them, which ought to then be trivial to distinguish from the others, but the reality is much more subtle than this. And second, cancer is often spoken about as though it is the same as having high blood pressure, or diabetes, or a sexually transmitted disease. But it isn't. Cancer is categorically different from all such conditions. In fact, it belongs in a category all its own. Cancer represents the inevitability of biological life to expire due to the imperfection of nature. It can be delayed with a good genome and by staying away from risk factors, but it remains an inevitability. Any kind of sweeping solution to this biological limitation would require complete control over the human genome, and the ability to edit the DNA of any cell at will, in order to reverse any mutations that may arise. This would be considered the realm of biotechnology, and we will spend the entire next chapter investigating such futuristic possibilities. For now, let's focus on what we do to treat cancer today.

Let's start with the least sophisticated option, radiation therapy. Just in the way that radiation from the sun can cause mutations in DNA, because of the high energy associated with ultraviolet light, we can focus even higher-energy light, such as X-rays or gamma rays, on particular cells with the intention of killing them. If a cell is ravaged with radiation, enormous numbers of mutations are sustained, and continued cellular function becomes impossible, so they die. Public perception of radiation therapy tends to be misled, typically because cancer is misunderstood. Cancer is just a bunch of cells that aren't supposed to be there. So to get rid of cancer, we have to kill cells. The trick is to figure out how to kill only cancer cells. Since this is such an overwhelming challenge, early solutions such as radiation didn't even attempt to differentiate. When a malignant tumor is highly localized in one spot, we can just blast that spot with radiation. There is nothing inherently sophisticated about it on the conceptual level, although certain technologies have become incredibly precise with this approach over the past decade. Either way, we simply try to kill cells. Inevitably, we kill lots of normal cells too, and this is what leads to the debilitating side effects of this treatment, which are not pretty. But if the prognosis is grim, the side effects are better than death, and if every last cancer cell is destroyed, then the cancer is gone, and the patient will probably live. Even when this approach was less refined, it was still a valid course of action as somewhat of a Hail Mary play. If the patient was likely to die, they had very little to lose by undergoing this treatment, and even less so now with the added precision. Of course, if metastasis has occurred, and the cancer cells are no longer in only one specific location, this approach is very unlikely to be effective, so let's check out some other options.

Getting a little more directed, there is chemotherapy. This is where specific drugs are administered that are able to recognize cells in the process of dividing. These drugs circulate around the body, killing dividing cells, and since cancer cells are rapidly dividing, the drugs are much more likely to kill cancer cells. Typically a combination of drugs is prescribed, each of which kills cells at a different stage of cell division, thus maximizing the cancer cells that will be killed. This is an appropriate course of action if the cancer is spreading, such that localized treatment will not be effective. The problem is that there are lots of normal cells that also divide quickly. Hair follicles rapidly divide, and this is why your hair grows so fast compared to the rest of your body. These will die, which explains why your hair falls out during chemotherapy. There are other cells in the blood and digestive tract that suffer the same fate, which again leads to nasty side effects, such as increased susceptibility to infection, weakness, and vomiting.

Beyond this, cancer treatment is an area of persistent study, and new strategies such as cancer immunotherapy are being innovated all the time. There are so many possibilities in terms of the recognition and destruction of cancer cells, which can target gene expression, DNA replication, or cell division, via any number of targets. This makes cancer an ongoing and dynamic aspect of modern medical research. The alt-health industry has a different take, however. It goes without saying that some dismiss any and all treatments from the medical establishment. Beyond this, there are those who make claims regarding so-called "superfoods," and allege that they have the ability to cure cancer. To be clear, this goes far beyond the notion that a particular food will reduce the risk of cancer, which is possible but only in very specific circumstances. We are talking about claims that specific foods can literally cure existing cancer. Broccoli, berries, kale, the wondrous and remarkable properties of these heroic foods are shouted from the proverbial alternative mountaintop. How do these foods achieve their therapeutic effect? What is the active ingredient, and what does it do mechanistically? Most outlets do not even attempt to answer these questions, because they reject this scientific line of discourse entirely, presuming that the reader does not know what cancer is. Others say that the foods flush out "cancer-causing chemicals," or something similarly vague. Some will mention "toxin removal," or reference other buzzwords we've discussed, which we know is an indicator of deception in this context. If an attempt is made to attach a scientist or medical professional to such a claim, no citation is offered to let the readers verify anything for themselves. It's just an empty argument from authority. And yet, all of this does not even begin to compare to still more questionable suggestions, such as yoga, tai chi, meditation, or ancient Chinese herbs. Some of these qualify as good exercise or an enriching practice, which can improve your quality of life, but the notion that they can have such specific anti-cancer therapeutic effects is beyond baseless. Not only are there no clinical demonstrations of these effects, no such demonstrations would ever be attempted, because the claims are specifically aimed at those who do not require scientific validity as a component to be persuaded. They offer only anecdotal evidence, which is the worst evidence possible, in any context.

*TL;DR—There is no diet that cures cancer.*

After knowing what we've learned about precisely what cancer is and what we can do about it, these alt-health solutions should seem appropriately ridiculous. What is it about these foods that could have any impact on any of the cellular processes we have outlined above? Beyond a light sprinkling of scientific terminology, why is no attempt made whatsoever to explain in a mechanistic way what these foods do in the body on the molecular level? The answer is simple. No such mechanism is known to exist. It is just another example of an appeal to the desire to view nature as mystical, benevolent, and healing. It is an actualization of the childish longing for Mother Nature to sing a sweet song and make the boo-boo go away. But she does not bend to our desires, beyond the best placebo our minds can conjure. Nature is the construct responsible for the fragility of the genetic material that resulted in the cancer in the first place, and nature does not have any quick fix for you. So we must not buy the narrative that such outlets are selling. No matter how much we want to believe that we can cure absolutely any ailment with what we find in the ground, it just isn't the case, and those who propagate such ridiculous notions endanger sick people by steering them away from legitimate treatment, only to end up chasing a mirage in the distance. Exercise and a healthy diet may keep certain conditions like diabetes at bay, and their importance is not to be minimized, but even in these cases, we must understand precisely how they correlate with broader health on a mechanistic level, rather than attempting to endow food with magical qualities.

> **TL;DR—Narratives that steer people away from legitimate medical treatment are dangerous and should be exposed as such.**

This is the key distinction that must be made if we are to discuss not generalized preventive measures, but actual treatment. Real treatments are drugs. Drugs are molecules, and they impact the body on the molecular level. That's how chemistry works, and that's how the human body works. Those who speak against this basic fact are purveyors of chemophobia, and nothing more. Because those who promote chemophobia must then steer clear of the chemical world, as it is what they blindly denounce, alt-health treatments tend to take the form of something familiar, like food, despite the irony that all food is an assemblage of chemicals. Food is delicious. There is a plethora of positive connotation to point to. This is just a side door into the mind, a salesman's script.

But if drugs are the answer, then why is it that so many people recoil upon a discussion of drugs, as though they are inherently evil? Many assign a negative connotation to this word, most likely because we so often discuss illicit drugs. These include recreational drugs that may or may not cause harm to the body, depending on the level of consumption, but are still viewed in a negative light by certain people. This also includes dangerous substances like some opioids that cause rapid addiction and can ruin a person's life. It's all the same word, because it's an umbrella term that describes an enormous panoply of substances. Drugs are not specifically unnatural, and they are not inherently bad or harmful. A drug is simply any substance that has a non-nutritional physiological effect when introduced to the body. So alcohol, marijuana, cocaine, aspirin, caffeine, nicotine, these are drugs. We consume them, with varying degrees of sound judgement, to have fun, loosen up, focus the mind, or get a buzz going. These goals are achieved on the basis of how these molecules interact with molecules in the body, whether by the inhibition of an enzyme, or the activation of a receptor, or any other mechanism.

For precisely the same reasons, pharmaceutical drugs are drugs. They serve a dramatically different physiological purpose, but they elicit some effect all the same, whether by the inhibition of an enzyme, or the activation of a receptor, or any other mechanism. And any treatment found in nature that has any legitimacy whatsoever contains some drug that operates by such a mechanism as well, whether it's the quinine in the bark of the Cinchona trees, or any other natural remedy that was stumbled upon over the millennia of human civilization that preceded modern science. Anyone who expresses the opinion that drugs are not a good way to cure disease is not to be trusted, because drugs are the primary way to cure disease, in whatever form they may take, natural or synthetic.

Alt-health advocates would reply to such claims with skepticism. A good diet and lots of exercise, that's all you need to stay healthy. To be fair, they're not all wrong. These are two absolutely crucial preventive measures that help us avoid certain conditions, such as heart disease, obesity, or diabetes. We can do a lot to minimize risk. But these measures do not make one immune to genetic variation. They do not regulate every bodily function. And they certainly have minimal influence on our vulnerability to certain pathogens. Pathogens are microorganisms that can cause disease. These most commonly include bacteria, which are the prokaryotic unicellular organisms we mentioned, and viruses, which are much tinier, and are actually acellular, meaning they are not made of cells, and are dramatically simpler in structure than a single cell.

**Bacteria** are absolutely everywhere. They're in the water, they're in the soil, they're in the air, and they're in you. There are more bacterial cells inside you than there are cells of your own. That may sound alarming, but don't worry, most don't do anything harmful, and some actually help you, as is exhibited by the symbiotic relationship we have with our intestinal flora, which we could not survive without. But there are certain bacterial species that exhibit virulence, which describes the ability to infect or damage a host. This is achieved by specific virulence factors, which are toxins produced by a particular bacterial species that interact negatively with components in the body and allow them to colonize. Most of the time, our immune system is equipped to handle these critters. But when bacterial infection becomes a problem, a common course of action is to utilize antibiotics. These are drugs that rely on the dramatic structural differences between bacterial cells and animal cells. Unlike the challenge with cancer, where our own cells are the problem, bacterial cells are totally different, and we can target certain structural aspects. For example, animal cells possess a plasma membrane. Bacterial cells possess both a plasma membrane and another barrier structure called a cell wall, which is made of a substance called peptidoglycan. Certain antibiotics called beta-lactams are effective against bacteria because they inhibit the bacterial enzyme that facilitates cell wall synthesis. If the bacteria can't build their cell walls, they can't survive, and our own cells are not affected whatsoever, because they lack this structural feature entirely.

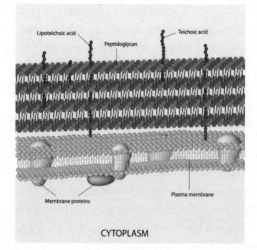

Bacteria possess cell walls in addition to plasma membranes, and this structure is the target of certain antibiotics, since it is not possessed by our own cells.

This strategy was first discovered through a study of nature. Certain molds produce a class of compounds called penicillins, which are beta-lactams, and they are effective anti-bacterial agents. Another class of beta-lactams are called cephalosporins, very similar in structure to penicillins and derived from molds. So there are other species in nature that came upon the problem of bacterial virulence, in this case molds, which are part of the fungi kingdom, being distinct from animals and plants. The chance production of these compounds was favored by natural selection as they armed the molds with the ability to ward off attack by certain bacterial species. We stumbled upon this fact in the early twentieth century, and it completely revolutionized the field of medicine. Suddenly we had a strategy to combat infection by microbes that was extremely effective and safe. Since then, we have tampered with the structures of these compounds to make deliberate modifications, dramatically enhancing their antimicrobial properties. This semi-synthetic approach is just one example of our ability to be inspired by nature and improve upon it.

Antibiotic resistance remains a concern, as bacteria live very briefly and multiply like the dickens. Bacterial enzymes also make a lot more mistakes during DNA replication than ours do, resulting in rapid mutation, and they can even participate in horizontal gene transfer, sharing genes from one bacterium to another without having to reproduce. All of this means that the accumulation of mutations required to sufficiently alter the bacterial genome so as to generate a new strain happens extremely quickly, much faster than evolution occurs for multicellular organisms. For this reason, it is possible for mutant proteins to be produced that no longer interact with our drugs in such a way that they are incapacitated by them. This is why we sometimes use multiple antibiotics at once, because when attacking multiple targets, it becomes increasingly improbable that a new strain could evolve resistance to all the drugs at once. It is also why patients must always finish their course of antibiotics, as every last bacterium must be accounted for, since it takes just one resistant mutant to allow a new resistant strain to proliferate.

**Viruses**, on the other hand, are different beasts. As we mentioned earlier, these are acellular. In truth, they are nothing more than some genetic material surrounded by a protein coat. They are just inert little genome pods. Technically speaking, viruses are not even alive. Living organisms are dynamic systems, with mind-boggling amounts of chemistry happening at all times. Metabolism occurs everywhere, with certain compounds being broken down and others being synthesized. ATP is produced and then spent elsewhere to produce motility, response to stimuli, and growth. Viruses do none of these things. They simply float around, and have proteins sticking out that can potentially be recognized by surface receptors on cells. If they are recognized, strictly by chance, the cell brings them inside, the viral coat dissolves, and the cell begins transcribing the viral genome, which can be either RNA or DNA. The resulting mRNAs are then translated to produce many new copies of the virus, instead of the cell focusing on its own needs. Often times, so many copies of the virus are made that the cell literally bursts, and the new viruses go to infect other cells.

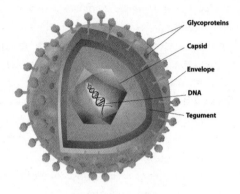

Glycoproteins

Capsid

Envelope

DNA

Tegument

An example of viral structure.

With viruses, treatment can be quite tricky. We typically rely on our own immune system, which is exceedingly complex, and equally astounding. A thorough summary of basic immunology is beyond the scope of this book, so we won't delve into many important details. But in so many words, there are immune cells that are able to tag foreign invaders, like viruses or bacteria, for destruction by other cells. In the case of a virus, in order for this to occur, certain immune cells need to interact with markers projecting from a virus-infected cell, which will include pieces of viral proteins. Once this interaction occurs, the body can then produce proteins called antibodies, which are able to recognize that marker, which is called an antigen. It can then tag the invader for destruction, along with any infected cells. The first problem is that the act of recognition takes place once cells are already infected. Furthermore, it takes several days for these antibodies to be produced, and in that time, a virus can replicate significantly. If the virus is eventually neutralized, but then enters the body again in the future, the defenses are already capable of recognizing the particular antigen from the previous exposure, which is why there are many pathogens that can't infect us a second time. But the first time it appears, it can be a huge problem.

A solution to this delay in immune activity has arrived by way of vaccines. With respect to viruses, these work as follows. We take some virus, and we generate what is called an attenuated version of the virus. This means an inactivated virus, or often just a piece of the virus, which includes the antigen that must be recognized in order to produce the corresponding antibodies. This inactive form, which is incapable of establishing infection, is introduced to the body. This allows the immune system to go through the motions of what we call the primary immune response, generating the necessary antibodies, without infection having to occur. Then, if the antigen is ever encountered by way of the associated pathogen, the immune system is already good to go, and will immediately enact the secondary immune response, which is the destructive portion of the response that will make quick work of the enemy.

This elegant solution is one of the most impressive achievements in the history of medicine, and its impact on human civilization can't be overstated. There are so many diseases of pathogenic origin that devastated humanity up until the twentieth century that are now almost completely unfamiliar to the general public in the twenty-first century.

Polio, smallpox, cholera, rabies, measles, mumps, these were ruthless killers, producing deaths in the millions annually. None of them concern the developed world any longer, thanks to vaccines. Smallpox, which has killed an estimated one billion people over human history, is now literally nonexistent. Even now as I write these words, I am quarantined in my home due to the coronavirus outbreak of 2020, nervously awaiting some sign of relent and a return to normality. Whatever the manner in which this problem will eventually be solved, it rests squarely in the hands of modern, science-based medicine. This is why anti-vaccine activists are thoroughly perplexing from a logical standpoint, and ultimately dangerous to our society.

Of all the anti-science sentiment we are discussing in this book, this particular example is the most damaging. It represents a complete failure of the public to appreciate the achievements of science, and the most egregious conflation of industry and malice. The modern incarnation of the anti-vaccine movement began on the basis of a singular study published in 1998 by someone named Andrew Wakefield, who had clear financial motives for his claim that the MMR vaccine, which inoculates against measles, mumps, and rubella, is linked to the onset of autism in children. When his results could not be reproduced, an investigation determined that he had been completely dishonest with his data and had conducted himself unscientifically. He was shunned from the scientific community, and the study was retracted. But the damage had been done. We are still feeling the impact of this singular fraudulent study today. Parents with autistic children looking for someone to blame got up on their soapboxes, amplifying this misinformation to the point of frenzy, and it is at the degree now where even mountains of well-articulated explanations from scientific authorities regarding the illegitimacy of Wakefield's study as well as genuine pleas regarding the safety and necessity of vaccination are falling on deaf ears. To some people, no amount of logic or reason can permeate the desperate desire to blame The Man and cling to a baseless narrative.

How can people be this unreasonable? Well, it's a pretty big ask for the common public to comprehend something as complex as autism, and the cause for its onset. It is immaculately clear now that autism has a genetic basis. Studies are ongoing, and point to interaction between many different genes, but the heritability of the condition is a clear indicator of the genetic basis, as is supported by twin studies and sibling studies. The increase in its diagnosis is simply a result of our very recent capacity to diagnose it, whereas in the past, autistic children were simply regarded as odd. Continuing studies are quite conclusive in this regard, and while there is a lot left to learn, there is no question that vaccines and autism have no correlation whatsoever. By what mechanism could any component of a vaccine induce such a thing in the first place? To the concerned sector of the public, the answer is simply "chemicals." Chemophobia rears its ugly head once more. Because this issue is so crucial, please indulge me for a moment.

Vaccines don't cause autism. Vaccines don't cause autism. Vaccines don't cause autism.

**TL;DR—Vaccines don't cause autism.**

I apologize for repeating myself, but it is effective, and this point must be made crystal clear. As for vaccine injuries, these exist but are rare. Individualized potential for them can be screened, and the minute risk is dramatically outweighed by the importance of maintaining the eradication of these diseases, a few of which are beginning to come back due to the growing popularity of anti-science sentiment. There are those who smugly insist that the unvaccinated can't possibly be a threat to the vaccinated. These individuals are undoubtedly mistaken, for several reasons. The first is that we have a necessity for what is called herd immunity. There are people who are unable to get vaccinated because they are immunocompromised in some way. This includes infants and people who have some kind of contraindication to vaccines, which means it could be harmful to them for a specific medical reason. The second reason, though it is less frequently discussed, involves the evolution of pathogens. Viruses, just as with bacteria, have genetic material that is susceptible to mutation. Mutation in the viral genome can lead to mutant viral proteins, including the proteins that are recognized by the host cell to allow for viral entry. If mutation yields structural alteration of these proteins, this can offer the virus a new strategy for entering host cells, or this can make it unrecognizable to the antibodies produced by the primary immune response upon vaccination. In other words, if a virus is given the opportunity to proliferate in a large enough host pool, mutation is inevitable, which can produce a new strain that the masses are not immune to. These are the main reasons why everyone who can get vaccines must do so. Some can't get them, but the virus must not have anywhere to go, or it will continue to thrive. The unvaccinated are objectively a massive public health risk, and they must be made aware of this fact, one way or another.

The danger of anti-vaccine activism is a superb example of what is at stake when combating ignorance. Human lives are in the balance. It is also a perfect example of how narratives lead us astray when applied in a sweeping manner. We know that anti-industry sentiment is rampant and sometimes legitimate. For example, climate change is an enormous threat to our way of life. Without coming anywhere near succumbing to alarmism, average temperatures are objectively increasing and sea levels are objectively rising, which puts coastal cities at risk. Carbon dioxide emissions are directly responsible for this trend, which is the single best example of legitimate anti-industry sentiment, as industrial practices are specifically to blame for this phenomenon. The greed of the few affects us all. It is perhaps forgivable to some extent that this anti-industry sentiment is adopted as mantra by so many, but it is misplaced to wield this worldview blindly, and baselessly apply it to other large industries, like industrialized medicine. Vaccines don't work simply because of what some industry says or wants. In truth, profits to be made on vaccines are quite negligible. Vaccines work because of science. They work because of painstaking ongoing research by hundreds of thousands of people in the fields of biochemistry, microbiology, human physiology, and immunology. So when we arrive at a particular narrative because of personal conviction, or even political affiliation, we should not be surprised when that narrative fails to apply to every relevant sector of our civilization. Take issue with the health care industry and the privatization of personal health. Take issue with pharmaceutical companies that offer kickbacks to doctors to overprescribe or even misprescribe a drug. Do not take issue with our study of the natural world and the minds that utilize this knowledge in order to solve universal problems with astounding ingenuity. Their work benefits us all and elevates our society one step closer to something we all should hope it can be.

> **TL;DR—Blindly applying anti-industry sentiment is never a good idea.**

There are those who would cite instances in which it was not just industry but also the science that specifically failed us. To be completely fair, I will mention one such major instance. In Europe, during the 1950s, a drug was developed called thalidomide. It was used to treat nausea in pregnant women, which is sometimes referred to as morning sickness. This compound, like many other organic compounds, exhibits a quality called chirality. This means that it is different from its mirror image. There is one so-called chiral center on the molecule, a particular carbon atom from which a portion of the molecule could project in one direction or the other. Depending on which direction, we get two different molecules that have precisely the same connectivity, but they differ in the way the atoms are distributed in space, such that they are mirror images of one another, just like your hands.

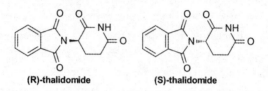

**(R)-thalidomide**          **(S)-thalidomide**

The structure of thalidomide, in both of its enantiomeric forms.

As we recall, proteins are also three-dimensional molecules, and the active site of a protein is a three-dimensional region that accommodates a particular molecule with a particular three-dimensional shape. What this means is that one version of thalidomide, which we can call one enantiomer, interacts with the target protein and does indeed achieve the desired physiological effect. However, by complete blind chance, the other version, or the other enantiomer, beyond simply being unable to interact with the target protein, happens to interact with some other protein or proteins. And in a totally unpredictable manner, this produces terrible birth defects. Thousands of babies were born severely malformed, suffering from a condition called phocomelia, which results in less than fully formed limbs, and nearly half of the babies died.

The issue was twofold. First, it was problematic that we did not know that the wrong enantiomer would do this terrible thing, and furthermore that the two enantiomers of thalidomide happen to interconvert in the body. To speak more generally, it was also an issue that we did not yet realize the importance of performing what we now call an enantioselective synthesis. This is a synthetic pathway that would lead us only to one enantiomer and not the other, rather than a racemic mixture, meaning half and half. This was, at the time, very difficult to do for reasons that require knowledge of organic chemistry to understand, but fortunately the lesson was not lost on the field, and we now place great importance on stereospecific synthesis to ensure that we always get the right version, or stereoisomer, of a particular chemical structure. We also now place great emphasis on reproductive toxicology, making sure to study the effects of a drug on an animal fetus, something that was not required when thalidomide went on the market.

But even still, much of the blame rests on industry. Although the effects of the drug on a developing fetus could not have been predicted without reproductive studies, even once these effects became clear, the drug was not immediately pulled from the market. This action did not come until there was intense pressure from the press and public. Selfish behavior like this is the reason the public distrusts the pharmaceutical industry. Not just because of ethical transgressions, but because of the long-lasting suspicion it elicits and the narratives that then propagate, like a bad personal reputation.

In the end, thalidomide is a cautionary tale. It is a reminder that we must never be cavalier as we forge ahead into unknown territory, and we must never let hubris get the best of us. But it is not justifiable to allow mistakes to convince us to abandon progress. If that were the case, every automobile accident would be a good enough reason to discontinue their use. Instead, we continue to innovate the automobile, to enhance its safety, in the eventual limit of removing the driver altogether, which will certainly reduce the number of accidents by an overwhelming proportion. In this way and many others, mankind is in a transitional phase. We are no longer at the complete mercy of nature, and we are not yet in its command. Huge advances in hygiene have done wonders in reducing mortality, namely by providing access to clean water and food to minimize pathogen exposure, as well as sanitation around the home and in medical procedures. This alone has had enormous impact on the average life span. But there is so much innovation that remains. Nature has provided our bodies with incredible safeguards from the external world, in the way of metabolic pathways that dispense of small foreign molecules, and antibodies that destroy pathogens, but even in this, nature reveals many foibles. Those same components of the immune system that kill invaders have the capacity to mistake our own cells for the enemy, resulting in a variety of autoimmune disorders that quite literally involve the body attacking itself. Nature is not a sentient, omnipotent deity. Nature is imperfect, and we are of nature. If we are to solve big problems, we must be bold. We must be brave. We must have vision. We must improve upon the cards that nature has dealt us. What does the future hold for the biological sciences? It's time that we discuss precisely this.

# CHAPTER 9

# Biotechnology and the Future of the Species

Think about the world as it exists today. Think about the technology you use on a daily basis. Your computer, your phone, your Netflix account. Now think of the world as it was ten years ago. If you're old enough, think about the world as it was twenty years ago, or thirty, or even forty. As you drift serenely into the past, notice how all the technologies that are commonplace in our modern lives start to fade away. Watch your cell phone lose its touch screen capabilities. Then watch it leave your pocket all together and jump back up on the wall next to your fridge, perhaps even back to a rotary dialing mechanism instead of buttons you can press. At the same time, watch the internet dwindle in relevance, its omnipresent marketplaces folding one by one, then returning to its original status as "just for nerds" and eventually disappearing entirely. It is suddenly no longer the case that all the music in the world exists for free on something the size of a candy bar. You can no longer watch any television program imaginable whenever you feel like it. No more knowing where you are and where you are going at all times, it's back to using paper maps. And perhaps most significantly, no more ceaseless interpersonal connection.

We don't have to go back all that far in order for technology, and by extension society, to become totally unrecognizable. I'm not that old, and I remember ancient relics like the rotary phone quite vividly. The rapid technological progress we've seen over a single generation is a firm indicator that this growth is exponential. For people who lived a few thousand years ago, human civilization did not change much at all in their lifetimes. The way it looked when they were born was pretty much how it looked when they died. Maybe someone would invent a slightly better shovel, or fishing tool, but nothing too revolutionary. Slowly but surely, we picked up steam. Whereas twelve thousand years separate the agricultural revolution from the industrial revolution, a mere two hundred separate the latter from the information revolution. It would seem that we are reaching the knee of the curve, where we may witness multiple revolutions of such magnitude within an individual lifetime. Already, technology becomes obsolete in a decade, or even less, depending on the context. What do you imagine the next few decades will bring? In what ways will we be shocked and challenged with the innovations that are soon to be incorporated into our daily lives? How will we build upon what exists today to create a better tomorrow?

Undoubtedly, computers will become more powerful. No one questions this prospect, and perhaps surprisingly, most do not seem too scared of it either. This is not the aspect of public perception that needs rehabilitating. Maybe this is because computers are inherently man-made. The manipulation and optimization of that which is born of mankind is not a challenging narrative. Improving upon the toaster oven doesn't invade anyone's worldview. It's the manipulation and optimization of that which is born of nature that seems to ruffle some feathers. It would appear that yet another echo of vitalism persists, suggesting that an attempt to improve upon nature is not just preposterous, but specifically unethical, and ill-advised. In short, public perception of biotechnology is not unanimously positive. Is this fair? Are people being closed-minded, or cautious? Does the modification of living organisms constitute hubris or progress?

Given the theme of this book, let's analyze the worldview of those who regard the entire field of biotechnology as unnatural. Of course, in a certain sense, they are correct. When we manipulate living organisms on the molecular level, we are enacting changes that nature would not have produced on her own, at least not on such an accelerated timeline. Sentience allows us to defy our own nature, first cognitively and now mechanically. But one could just as easily argue that when organisms which nature produced engage in self-improvement, this is essentially just an extension of nature, another natural process unto itself. After all, we are natural, and we use our understanding of the natural world and the materials that nature affords us in order to build new constructs. Are these not then natural in a certain sense?

Perhaps it is best not to get bogged down with semantics. More importantly, we ought to avoid the temptation to simply fear what is new and unfamiliar. People have a tendency to accept all of the technology that exists at the time of their youth, and fear all the advancement they are exposed to later in their lifetime. But what would people from just a few centuries ago think of what we have today? Cars, airplanes, television, skyscrapers, these would be enough to place them in a fear-induced coma. Similarly, in a few more centuries, people will find their technology to be just as mundane as our smartphones are to us, and they will look upon our reality as ancient and primitive. That is the way of things. That is the way of progress. It is best to try to view the path of our species in a nontemporal way. From a certain point of view, there is not that which came before and that which will come to be. There simply is. It is a mental exercise that may allow us to examine biotechnology in a more level-headed way.

> **TL;DR—Let's not fear things solely because they are unfamiliar.**

All of this talk of progress does not mean we should be cavalier in our tinkering. Innovation should be done prudently and ethically, insofar as we can establish ethical guidelines for that which does not yet exist. But when there is backlash against a particular innovation, ignorance is not a valid argument, so let's hit some key points. There are two main sectors in which biotechnology is paving the way to the future. Those are medicine and agriculture. Let's touch on medicine to start, as it is the best realm in which to introduce the concept of genetic engineering.

We employ certain microorganisms in biochemical reactors to help us produce biomolecules. For example, if there is a particular protein of interest to humans, we will typically identify the corresponding gene that is expressed to produce it. Then we can either isolate that gene or synthesize it ourselves, and insert it within the genome of an *E. coli* bacterium. We can now refer to this bacterial genome as **recombinant DNA**, because it contains DNA from more than one organism.

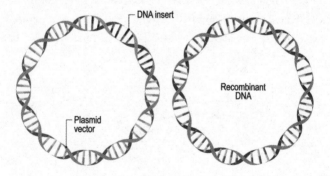

A gene can be inserted into a circular DNA molecule called a plasmid, which is part of a bacterial genome, and the bacterium will express that gene.

This bacterium will then express the genes in its genome, including the foreign gene, and it will also divide rapidly to produce many more bacteria at an exponential rate. Within a

day or two there should be countless numbers of bacteria, all
of which are producing the protein of interest, like good little
factory workers. We can then harvest the protein and utilize
it as we see fit. One application of this technique is the large-
scale production of insulin, an important peptide hormone that
regulates levels of glucose in your blood. An inability to produce
insulin is the primary cause of type 1 diabetes, so regular
injections of insulin will help avoid complications associated
with insulin deficiency, and we can use bacteria to make all the
insulin we want.

Seems like a slam dunk for science, right? Well, what if
we go beyond implementing a single gene, what about
the engineering of completely new organisms with novel
functionality? We can do that too, and this may be one creative
strategy for combating climate change. What if we were to
engineer microorganisms or plants that could consume carbon
dioxide at many times the normal rate, thus effectively removing
it from the environment and restoring optimal atmospheric
conditions? This sounds fantastic, of course, and is more feasible
than many might think. But do we have the right to do this?
Who says we can tinker with DNA and play God in this manner?

At this point one thing must be made abundantly clear. It is
not the case that we have been living innocently in the pristine
splendor of nature, never tampering with its wonders until
the arrogance of twentieth century industry. We have been
manipulating nature for all of recorded history. When we
use microorganisms for certain tasks, like we do with yeast

in food and alcoholic beverage production, that qualifies as biotechnology. Forcing microbes to brew our beer is not any different than forcing them to synthesize a particular protein, at least from a philosophical standpoint. They did not consent then, and they don't consent now. But unicellular organisms are not conscious, so there is no basis for the suggestion that they would protest, when they are simply performing the chemistry that they inherently perform.

When we engage in the breeding of domestic animals, like dogs, again we are interfering with natural processes. We are guiding the evolution of living organisms based on a set of aesthetic criteria that we, not nature, have selected. We make the distinction by labeling this process artificial selection, rather than natural selection, which is based exclusively on survival. Nature gave us wolves. Tiny dogs that fit in women's purses therefore exist only because we frivolously decided that they should, not because of natural principles.

Dog breeding is an example of artificial selection.

The same can be said for virtually everything in the produce section at the grocery store. Many believe that those foods have existed that way for millions of years, as though we were born into a Garden of Eden with perfect fruits and vegetables dangling from vines and trees. This simply isn't the case. Those foods look and taste that way because we made them that way, again through artificial selection. Many of them are literally of our own invention. Broccoli, cauliflower, kale, cabbage, Brussels sprouts, these are all derived from another plant, *Brassica oleracea*, which was not particularly delicious. An overwhelming proportion of our favorite vegetables are not naturally occurring, but came about by identifying specific traits and breeding selectively for centuries, just like we did with dogs. What gives us the right to do that? If one is to make the claim that we do not have the right to tamper with nature, then what are we doing breeding dogs? Why are we allowed to invent new plants?

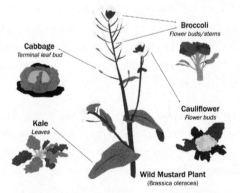

Many kinds of vegetables we are familiar with today were derived from the wild mustard plant, *Brassica oleracea*.

> **TL;DR—Manipulating nature is not new, we have
> been doing it for millennia.**

There are those who insist that the distinction is made on the mechanical level. That once we use machines to alter DNA, that's when the line is being crossed. But why? It's quite an arbitrary place to draw the line. Whether guiding the gene pool by selective breeding, thereby altering genetic structure, or by more directly altering genetic structure in a mechanical way, there isn't an explicit difference. The only real reason people draw the line in this place is because dog breeding is readily comprehensible, while the practical details of genetic engineering are not. This doesn't make them any different fundamentally, it is just the complexity of the technique that is different. We are still looking at the traits of biological organisms and guiding their proliferation, we simply avoid many physical and temporal limitations by manipulating on the genetic level rather than the organismal level. The fear of the unfamiliar rears its ugly head once more.

Let's look to the agricultural sector for further examples. We rely on crops for food, and farmers have always faced tremendous challenges in the way of pests. These are organisms that cause damage to crops, which are typically either insects that eat the crops, or other plants like weeds, which compete for water and nutrients, thus reducing crop yield and quality. In the past, when harvests were compromised, people would die. It was such a huge problem that even biblical scripture is full of tales of locusts and other such pests. Nowadays, with nearly eight billion people in the world, great care must be taken to maintain expected crop yields, or large-scale famine results. Early solutions involved pesticides and insecticides, which actually first came into practice thousands of years ago. In a more modern context, we may have heard of popular insecticides such as dichlorodiphenyltricholorethane, or DDT, which came into use in the 1940s. At first, this compound appeared to be safe to use, and it did dramatically reduce cases of insect-borne diseases like malaria and yellow fever. But detrimental environmental and health effects soon began to surface, and opposition toward industrial use of this compound was a focal point of Rachel Carson's book *Silent Spring*. This was essentially the birth of the environmental movement and the birth of modern chemophobia, admittedly with firm basis.

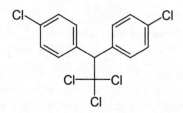

The structure of dichlorodiphenyltrichloroethane, or DDT.

As time went by, better strategies were implemented. Compounds with better environmental and toxicological profiles were identified, which could be used with greater specificity and in smaller amounts, along with integrated pest management systems that further reduced the necessary pesticide usage. Eventually genetic engineering was applied to the problem. It became possible to insert a gene into a plant's genome so as to bestow a particular crop with innate resistance to pests, or to a particular herbicide, making it much easier to use that substance to kill weeds without destroying crops. Plants can be engineered to cope with harsh conditions. They can even be engineered to include additional nutrients, such as the famous "golden rice," which is capable of synthesizing beta-carotene, a precursor of vitamin A. This is tremendous in combating vitamin A deficiency, which kills hundreds of thousands of children every year.

We won't get too far into the weeds on this, so to speak, because it is a tricky subject, and doing justice to the topic of genetic engineering in agriculture requires a book all to itself. To speak briefly and generally, anti-GMO sentiment is not completely without basis. The environmental consequences of large-scale manipulation affect many species, and the ecological impact is not always properly studied before such modifications are introduced. However, popular negative sentiment is almost always misdirected. The scrutiny that should be applied to agricultural practices and their consequences does not rest solely on the presence or absence of pesticides, and not whatsoever on the natural or synthetic status of their origin. Soil erosion, overfishing, clean water depletion, these are the issues that relate to ethics and the longevity of our food supply, not the narratives promoted by peddlers of chemophobia.

As we have come to repeatedly reinforce as the central theme of this book, science and industry are two separate realms. Genetic engineering is a good scientific idea. It could solve enormous problems, and in a way that is specifically optimal to both the environment and our health, as ironic as that would seem to the activists who simply refuse to learn the science behind it. When these tools are misused in industry, we must investigate their products and practices on a case-by-case basis, so that we may collectively react in sound, informed fashion. That is how legislation gets passed, how regulations are implemented, and how true progress is made. Industry must continue to be regulated, and perhaps more strictly than it is now. But it must not be handcuffed, or human civilization suffers.

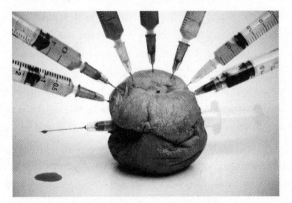

An example of alt-health propaganda.

However, an educated, calculated response is not what we see. When it comes to opposition toward biotechnology in agriculture, we are typically dealing with ridiculous propaganda in the form of fear-mongering images like the one above. Here we see a disgusting-looking piece of fruit stuck with many syringes containing various unidentified colorful fluids. Honestly, what is this? What is the intended implication? It has absolutely nothing to do with genetic engineering whatsoever. Who is injecting fruit with anything? What could these fluids be in the first place? This is nothing more than alt-health absurdity, which serves only to instill the irrational fear of "nasty chemicals in here." You should be personally offended that anyone on earth thinks this could work in steering the consumer habits of any sane individual, and even more disappointed that it actually does.

The result of alt-health propaganda surrounding agriculture is the organic craze. Earlier we mentioned the definition of the word **organic** in a scientific context. It means carbon-based. Simple enough, right? What does organic mean to people who insist on aligning their grocery purchases with this term? I guarantee that ninety-nine out of a hundred such people could not muster a logically coherent response to this question. The notion that it means "chemical-free" is absurd, as everything we eat is nothing but chemicals. The notion that it implies a lack of pesticide usage is objectively false. All crops are subject to pests, and all farmers take measures to protect their crops, as do the plants themselves, given that almost all pesticides in use are compounds that plants naturally produce to defend themselves. To those who place nature on a pedestal, organic means that the pesticides which are used are not synthetically derived. But to recapitulate prior chapters, who cares whether a pesticide is natural in origin, and especially whether naturally derived or synthetic in production? It is completely irrelevant, and severely misguided in this context, since the pesticides plants naturally produce are just as harmful as synthetic ones, if not more, and present in dramatically greater quantities than any synthetic pesticide residue that can be found in virtually any produce we eat. I hate to burst any bubbles, but organic farming does utilize pesticides, many of which are highly toxic, and often not thoroughly tested.

Because "organic" is little more than a buzzword, an organic label does not actually convey any relevant or useful information.

The best thing one can possibly say about organic produce is that it is sometimes grown in smaller yields with greater care and attention paid. If you buy organic produce for the higher quality, then there is no issue. But that's not the story we are sold. We are told a tale of Big Business with their nasty chemicals, and the Robin Hoods of the soil that keep us safe and healthy. That's why the term "organic" is essentially meaningless in this context. Its definition is vague and universally misunderstood, which relegates it to the status of a buzzword meant to make consumers feel warm and fuzzy inside. An organic label is just a sticker that allows people to feel like they have fulfilled their social justice quota without truly having done anything.

> **TL;DR—With produce, the term "organic" is largely just a buzzword to drive sales.**

This type of baseless virtue-signaling is boundless. There are actual products available for purchase, such as salt and baking soda, that boast GMO-free status. Just what, pray tell, is non-GMO salt? Salt is not a living organism. It does not have genes. Therefore, there are no genes to modify. So what in the blazes is this label doing on this product? Technically it isn't a lie, but it is complete madness. The degree to which people fear having science in their food drives them to such bizarre consumer habits that we can barely find anything on the shelves that doesn't prioritize this kind of message, whether it makes any sense or not. And when brands reinforce this fear, it's essentially a form of gaslighting. It's a quiet whisper in the ear, reassuring the consumer that everyone is out to get us but them, and if we just keep buying what they sell, everything will be ok. Is there really a difference, therefore, between the greed of the alt-health industry, and the greed of the industries it criticizes? Absolutely not. And the fact that a capitalistic enterprise would wave a green flag to disguise its profit-obsessed intent as altruism is absolutely disgusting. Images of attractive people surrounded by radiant fruits and vegetables while meditating is a marketing tactic like any other. It's the sale of an identity. It's no different than an advertisement for a handbag that purportedly embodies glamour, or a soft drink that is somehow specifically for doing extreme sports. Just the way the fashion industry tells you how you should want to look, the alt-health industry tells you who you should want to be. There is nothing wrong with wanting to be fashionable or healthy, but we should strive to resist such a suggestible state of mind, whether in the context of this or any other transparent marketing ploy.

Organic food products are marketed in precisely the same manner as beauty products or high fashion, by selling the consumer an identity.

Let's now return to the medical realm for a dose of optimism. What kinds of techniques outline the current forefront? One fascinating example is gene therapy. There are a number of disorders that can be traced to a single defective gene, such as cystic fibrosis and hemophilia. Some mutation has arisen, which alters the product of gene expression, and the resulting protein does not perform its function as intended, which creates problems for that cell, and by extension, the organism. We could try to mediate this issue with drugs, but what if instead we could fix this gene? That would necessarily solve the problem for that cell, and if this could somehow be done for enough of the cells that possess the mutation, or more preventively at the stem cell level, prior to organismal development, it would solve the problem for the organism, definitively curing the disease. That is precisely what gene therapy seeks to do. Take for example a type of severe combined immunodeficiency that causes bone marrow cells to be unable to produce a vital enzyme, an issue which stems from a single gene. Because bone marrow cells include stem cells, which are undifferentiated cells that give rise to all the cells in the blood and immune system, this can be a huge problem. A solution to this is as follows. We can synthesize an RNA version of the gene of interest, and insert it into a retrovirus. A retrovirus is a type of virus that has the ability to generate a DNA transcript of its RNA genome, which it then inserts into a host cell for replication. We allow this retrovirus, containing our cloned gene, to infect bone marrow cells that have been removed from the patient. The virus is taken into these cells, and viral DNA, containing the normal version of the gene of interest, is inserted into the genome. These recombinant cells are then injected back into the bone marrow of the patient, and as these continually divide over an extended period of time, more and more cells will have the capacity to produce the vital enzyme, and the disorder is alleviated.

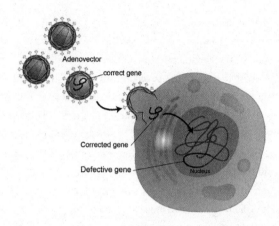

In gene therapy, retroviruses are employed to insert a specific gene into
cells that possess a mutated version of that gene.

Many detractors raise an ethical concern. Is it appropriate to
modify the genome of a living human? Well, it is worth noting
that this has already been done through blood transfusion
and organ transplantation. These both introduce living cells
with foreign DNA into someone's body. Is gene therapy really
that different? Furthermore, retroviruses exist in nature. It may
come as a surprise, but around 8 percent of human DNA is of
nonhuman origin, incorporated into the genome by retroviruses.
So this is actually a tactic applied by nature, not just some wild
human invention. Objections regarding cost are obviously much
more reasonable, but this is a burgeoning technology. With
innovation and refinement inevitably comes reduction in cost, as
can be seen countless times in history.

There is an even more advanced and site-specific genome editing technique, which is called CRISPR-Cas9. This will be more than a little technical to describe, so we will do so in the briefest manner possible, just to be thorough. CRISPR stands for clustered regularly interspaced short palindromic repeats, and Cas9 stands for CRISPR-associated protein 9. That's a mouthful, so let's unpack it a bit. CRISPR-Cas9 is an antiviral defense system found in certain bacteria. Specifically, Cas-9 is a type of enzyme called a nuclease, which means it is able to cut a DNA strand, and in bacteria it is directed toward viral DNA as a safeguard. The nuclease is directed to its target by a piece of guide RNA which interacts with the nuclease and allows it to snip only when a specific sequence of bases is recognized. We can modify this strategy to engineer nuclease enzymes that are able to cut not just viral DNA, but any DNA sequence of our choosing. This is done by modifying the sequence of bases in the guide RNA, which allows the enzyme to anneal in highly site-specific fashion. In other words, this is a method of removing or modifying highly specific sequences of DNA *in vivo*, which means inside a living organism. The advantage of this approach over gene therapy is that with a retrovirus it can be difficult to control where the gene is inserted. It is somewhat random, which poses an issue. Instead here, we can snip a faulty gene with great specificity and insert a healthy copy, offering endless medical applications. We can even insert novel genes into embryos to produce transgenic animals, which are animals that possess one or more genes from another animal. This technique is fast, cheap, and accurate, and will only grow increasingly so with time, so we can reasonably imagine that it will one day be used to treat humans in a variety of ways.

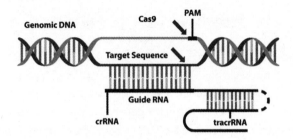

Visualizing the CRISPR-Cas9 technique for genome editing.

Fundamentally, whether we are discussing retroviruses or CRISPR is irrelevant. When addressing naysayers of genetic engineering as a concept, the important question is, who says we can't modify our genomes? In what gospel is it written that we may improve ourselves socially and morally, but never biologically? Of course one could argue that it is a slippery slope. If technology is indeed used to genetically engineer humans in sophisticated ways, what should be the guidelines? This issue could potentially lead to the practice of eugenics, whereby efforts are made to control the genomes of a population. This has been disastrous in the past, and under the wrong influence, could be disastrous in the future. We are identifying yet another reason to strive for a scientifically informed populace. We must not allow ourselves to be manipulated by political powers on the basis of ignorance.

Fear is the best tool for getting a citizenry to fall into line, and ignorance is the easiest way to keep a people fearful.

> **TL;DR—Genetic engineering is inevitable,**
> **so learn about it now.**

Let's highlight in rapid-fire a few more fascinating innovations we may see implemented in the coming decades. On the livestock front, things are a bit of a mess. Conditions for animals in factory farms are abominable, and their sheer numbers have environmental ramifications, given the emission of methane and nitrous oxide. Excessive proportions of land and agricultural output go to housing and feeding livestock. There is also some argument regarding the ethics of slaughtering and eating animals in general. Once again, like a brave knight, biotechnology has the potential to swoop in and solve this host of problems with one stroke of its sword.

Clean meat is a popular way of referring to laboratory-grown meat. We are not talking about plant-based meat alternatives here. This is actual animal tissue that is cultured and grown without necessitating the entire living organism. Every single environmental and ethical concern associated with eating meat instantly vanishes. If there is no animal, then there is nothing that is being harmed or slaughtered. We could eat a burger that was never a cow, or a chicken wing that was never a chicken. Beyond this, the tissue requires dramatically fewer nutrients than a living organism and produces no waste. Furthermore, the meat is not some substitute, so there is nothing about taste or texture that is being sacrificed. Believe it or not, as much as this sounds like science fiction, it is already a reality. Clean meat is currently being produced. Big challenges remain, including scaling up production, as well as bringing costs down, which remain astronomical for the moment. But these challenges will almost certainly be met, and we will probably see clean meat as an affordable option in restaurants and supermarkets within a decade. Imagine a reality where we can eat cheeseburgers, steak, chicken, and all of our other favorites without raising and killing a single animal. It all sounds so surreal that these practices produce detractors on that basis alone. But the production of these cell cultures proceeds by the same processes as what occurs inside a living organism, just without having to be inside of one. It is not "Frankenfood," as many would accuse, and there are myriad aspects of current meat production that are terribly unnatural anyway, if that sort of thing matters to a consumer. But we shall see how this matter unfolds.

Going back to the medical realm, among the most exciting emerging fields is that of nanorobotics. We will soon be able to produce robots with components on the approximate scale of nanometers, which are billionths of a meter, hence the term nanobots. Because of their size, they would be able to navigate the human body with ease, and with endless applications. These could be used to deliver drugs to their intended target, instead of requiring that drugs saturate the bloodstream, thereby going everywhere in the body. This would enhance efficacy and minimize side effects. Nanobots could also monitor a wide variety of bodily functions and alert the host if anything irregular is detected, thus bringing attention to an issue well before physical symptoms become apparent. These tiny machines could patrol the body, repairing damage and killing pathogens, or even cancer cells. This would completely replace approaches like chemotherapy, as nanobots could identify cancer cells without necessitating any mechanisms of surface protein recognition. Visual confirmation could potentially be enough, particularly if guided by the judgement of a physician. The possibilities are endless, and it is possible that in a century or so, healthcare will be dominated by tiny machines. Of course, there are many questions. What if these nanobots make errors? What if they are hacked somehow, and reprogrammed to damage the host? With a concept so futuristic, it is often difficult to spot the line between paranoia and prudence. But as long as we are educated and maintain an open dialogue, we will find our way.

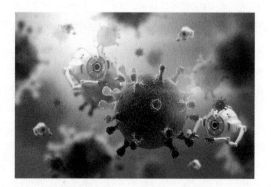

Nanobots could be the future of medicine.

Finally, let's get back to the thread that was initiated in Chapter Five. How do humans die? Let's say that futuristic nanobots are able to destroy all the arterial plaque, kill all the cancer cells on sight, and maintain our bodies in optimal health like the perfect little bodyguards we hope they will be. Humans would still die of old age. But can this issue be addressed? Amazingly, even this problem is not outside of the scope of biotechnology. As it happens, aging is not some biological inevitability. It is actually a rather arbitrary cellular process, and our understanding of how this works on the molecular level has grown quite sophisticated.

One component of aging is associated with a limitation in DNA replication. Every time a chromosome is replicated, DNA polymerase can only move in one specific direction along a strand. For one strand, the leading strand, this is done continuously. But on the other, the lagging strand, replication must be done one chunk at a time. And once the end of the strand is reached, there will always be a small portion that can't be replicated. This means that every time DNA replication occurs, a chromosome will become slightly shorter. Luckily, we have a safeguard in the way of sections of DNA at the ends of every chromosome called telomeres. These are sections without any genes that act as a buffer, and furthermore an enzyme named telomerase will regularly extend these sections so as to maintain them despite repeated shortening. But over time, this is not enough to prevent the shortening process from eating through the telomeres, and eventually causing loss of DNA within actual genes. This renders a cell unable to produce particular proteins, which causes all kinds of problems, and we refer to this state as senescence. But we have found that it is possible to selectively destroy senescent cells in mice by targeting them for apoptosis, which prevents issues like organ failure that are associated with old age. Other approaches that may curb the aging process involve stem cells, stimulating cellular ability to recycle certain components, or even countering specific genetic predisposition toward enacting the aging process. The key is to realize that aging is not mandatory. There are species that do not senesce, and therefore do not age, like *Hydra*, a genus of tiny freshwater organisms. Indeed, nature does not administer this bitter medicine to all of its creatures, as much as we paint this as a fundamental inevitability. Having already successfully expanded the life span of mice to a time and a half, and nematodes seven or eightfold, it would seem that the secrets are already being unraveled.

We may see anti-aging treatments administered on humans in the near future, and this offers perhaps the trickiest ethical terrain that we will discuss. Let's say we develop the capacity to end aging altogether. Who gets the treatment? Only the rich, or everyone? For those who are so lucky, how long shall we live? Do we implement a legal life limit, or do we leave room for immortality? If people begin living indefinitely, how can we continue to bring about social change? Our society depends on a constant recycling of power and ideas, how will this be possible if people never die? It is questions like these that make even a staunch optimist such as myself quiver with unease. But we may have to answer them within our lifetime. Are we truly at the knee of the curve? Will the twenty-first century bring about such explosive change that its continuation will be in the hands of artificial intelligence, leaving humans as mere witnesses to the evolution of sentience on this planet? Will we instead merge with technology and become *Homo deus*, the next species in an evolutionary chain that will be capable of colonizing the galaxy and beyond? These questions border on the philosophical, and they beg for an understanding of the fundamental physical nature of the universe, so let's move forward and see if we can put a dent in that as well.

# CHAPTER 10

# **Energy Defined**

There is a particular word that possesses a great many connotations. That word is **energy**. The concept of energy, as indispensable as it is to the domain of science, is terribly misunderstood by the general public, precisely due to the variety of contexts in which the word is used. Consider the following statements:

*I just don't have much energy today.*
*Society needs to start transitioning toward renewable energy.*
*This room has such good energy.*
*Astrophysicists are researching dark energy.*
*Cells use ATP as a source of energy.*

How can one word be used so many ways? Are some of these uses metaphorical, or can they all be taken completely literally? How may we unpack all of this to understand precisely what energy is? Well just as with previous chapters, we need to do a little bit of homework. We've already seen how significantly a bit of effort helped us understand the molecular world, so we need to do the same for some basic concepts in physics, because physics is the field of science that concerns itself with the most fundamental nature of the universe. Don't worry, the payoff will be substantial, so let's get started.

The logical way to begin is with the most rigid textbook definition of energy. This will require that we remove any and all existing knowledge we think we have about this word, so that we can purge ourselves of all the extraneous connotations it has come to possess. Now that you're working with a blank slate, here it comes. *Energy is the capacity to do work.* This single sentence completely defines the term. Unfortunately, we must now define work, because again, in common parlance this word is used to describe professions, actions, locations, and so many different things. In science, **work** refers to action done on an object whereby an applied force causes a displacement of that object. In other words, when you apply a force on an object and it moves, such as when you push something across the floor, you are doing work. This also means that if you apply a force on an object and it does not move, you are not doing work. So although pushing as hard as you can against a brick wall will make you very tired, you are technically not doing any work on the wall. This may seem counterintuitive, which is why we need to get rid of our prior conceptions regarding these basic terms. If work means moving something, and energy is the capacity to do work, then energy is the property that is responsible for the motion of any and all objects.

*TL;DR—Energy is the capacity to do work.*

Let's quickly talk about units of measurement associated with force, work, and energy, as this can help concretize these concepts. Forces are measured in Newtons, which are named after the famous physicist, Isaac Newton. One Newton, or 1 N, is defined as one kilogram times one meter per second squared, or 1 kg m/s². This means that when a force of 1 N is applied, it will cause a one kilogram mass to accelerate at 1 m/s². If this still seems confusing, don't worry, let's briefly define the terms position, velocity, and acceleration. Position communicates where something is located, like perhaps your position on some number line, where we arbitrarily designate motion to the right as the positive direction, and motion to the left as the negative direction. Velocity means change in position over time. So if you are moving to the right, and you travel one meter for every second that elapses, your velocity would be one meter per second, or 1 m/s. Now just as velocity means change in position over time, acceleration means change in velocity over time. So let's say you are standing still, and then you suddenly start moving to the right, but you get faster as you go. After one second has elapsed, you are moving 1 m/s. But you are speeding up, and after two seconds have elapsed, you are now moving 2 m/s. After three seconds, you are moving 3 m/s, continuing to move faster and faster as more time passes. Because your velocity increases by one meter per second for every second that elapses, your acceleration is 1 meter per second per second, which we can write as (1 m/s)/s, and when we divide a fraction by something else, we can just combine the denominators, so we multiply seconds together to get 1 m/s².

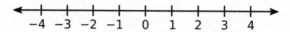

We can use a number line to help imagine position, velocity, and acceleration.

Only forces can produce accelerations. If an object is accelerating, which means its velocity is changing, whether speeding up or slowing down, there absolutely must be some force acting upon that object which is producing that change in velocity. This fact describes one of Newton's laws of motion, which we will discuss in more detail in the next chapter.

With these definitions now clear, we can see that a force of 1 N will produce an acceleration of 1 m/s² on a 1 kg object for as long as that force is being applied, such as when actively pushing something, or when pressing your foot on the gas pedal in a car. Work, on the other hand, is defined as force times distance, and this will be measured in joules. Joules are defined as Newtons times meters, or Newton-meters, so if a force of one Newton is continuously applied over a distance of one meter, one joule of work has been done. Energy is also measured in joules, which is why an object in motion is able to do work on another object. Ultimately, energy is a property of an object or system that can be transferred into another object or system, typically in the form of work. We must understand that energy is not an object, or a substance, or an entity of any kind, despite colloquial usage of the term.

*TL;DR—Energy is not a substance that can be possessed by an object, it is a property that is transferred from one object to another.*

While that is the bottom line regarding the definitions of these terms, this may all still be a bit abstract and confusing, so let's go over some different types of energy, and everything ought to become clearer. First up is kinetic energy. This is the energy an object possesses by virtue of its motion, and it is defined as one half of the mass of the object times the square of its velocity. So a greater kinetic energy is associated with a greater mass or a greater velocity, as can be seen with a slow-moving but enormous cruise ship, or a tiny but speeding bullet. Conversely, there is potential energy. This is the energy an object possesses by virtue of its position in a field. This field would most commonly be a gravitational field or an electromagnetic field. To elucidate this, imagine lifting a ball upwards, away from the ground. In doing so, you are increasing the potential energy of the ball, because it has the potential to fall a greater distance down toward the ground, due to the gravitational field generated by the earth. It may seem strange that an object can possess more or less energy simply because of its location, but think of a compressed spring, which exhibits elastic potential energy. The spring has the potential to expand if released, and we can feel that potential pushing outwards. In the same way, the ball has the potential to fall if you let go. When you then drop the ball, the potential energy of location is converted into kinetic energy of motion, as it moves some distance until it hits the ground. This is an illustration of conservation of energy, whereby energy is converted from one form to another, in this case from potential to kinetic, but the total energy is conserved. When the ball hits the ground, the kinetic energy it possessed while falling is converted into still other forms, such as thermal energy, so let's continue down the list of the types of energy.

Understanding thermal energy, or heat energy, will require that we define the term temperature. We have all heard this word before, as it is used every day to describe the weather. But what does it really mean, and what does it measure? Temperature is a word that refers to the average kinetic energy of the particles in a system. So hotter systems have faster-moving particles, and colder systems have slower-moving particles. When it is hot outside, atmospheric molecules, which are primarily nitrogen and oxygen, are moving faster, on average, than they are on a cold day. We experience that hotter temperature when those atmospheric molecules collide with the molecules in our skin, producing a particular sensation, which we perceive as "hot."

So we must understand that what we refer to as heat is actually just a transfer of kinetic energy from one object or system to another. The object or system with a greater average kinetic energy for its particles is the hotter one, and the object or system with a lower average kinetic energy for its particles is the colder one. When these two systems come into contact with one another, the particles in one system collide with the particles in the other, or vibrate against one another if the objects are solid, and kinetic energy is transferred by virtue of these interactions. Therefore, an object feels hot to the touch if its particles possess a lot more kinetic energy than the particles in your hand, and will therefore transfer kinetic energy into your hand. An object feels cold to the touch if its particles possess a lot less kinetic energy than the particles in your hand, which means your hand will transfer kinetic energy into that object. So perhaps counterintuitively, objects cannot "contain" heat, as heat is not a substance, or an essence, or technically even a type of energy. It is a word that describes a transfer of kinetic energy. Heat is energy in transit.

> **TL;DR—Heat is just a transfer of kinetic energy from faster-moving things to slower-moving things.**

Next up is chemical energy. Just as thermal energy is really a type of kinetic energy, chemical energy is a type of potential energy. In the same way that a ball has the potential to fall to the ground, a molecule has some potential to undergo a chemical reaction, and this potential is determined by the types of bonds in the substance. If a chemical process is spontaneous, it is said to be exergonic, and chemical energy will be released. If the process is nonspontaneous, it is endergonic, and chemical energy will not be released. But in short, chemical energy is the potential energy associated with the structural arrangement of atoms within a molecule. Some reactions must absorb heat from the surroundings in order to occur, and some reactions release heat to the surroundings, so this is another example of conservation of energy.

Finally, the most abstract form of energy we will discuss is matter. This includes water, air, your desk, this book, your body, and everything else that has mass. Matter itself is an ultra-dense form of energy, according to the principle of mass-energy equivalence, as is described by the most famous scientific equation of all time, Albert Einstein's $E = mc^2$. Here, E is energy, m is mass, and c is the speed of light. Light is extremely fast, the fastest thing there is, so c is a huge number. It is also squared in the equation, which makes it bigger still. What this means is that a teeny tiny amount of mass contains an unbelievable amount of energy. We will elucidate this further when we examine nuclear energy.

*TL;DR—Matter itself is a form of energy.*

Now that we have a reasonable understanding of what energy truly is, it is time to dissect the common usage of this word. Revisiting those five statements at the beginning of the chapter, do they now make sense? The one about ATP we already went over in discussing biochemistry and biology, and this should make even more sense after having discussed potential energy, as we can now understand the chemical energy that ATP and ADP possess. The one about not having energy today, that could be considered an extension of the one about ATP, although it is used a bit liberally, and non-literally, to refer to a state of exhaustion. Dark energy is too esoteric to get into here, and renewable energy will be described a bit later. For now, let's focus on the statement regarding a particular room having "good energy." When the word is used in this context, it is entirely metaphorical. It means that one gets a good impression from that room. The same can be said when it is applied to a person, or any other object. It is totally acceptable to use the word this way, provided that it is acknowledged that the usage is strictly metaphorical.

But there are those who use energy in a metaphorical context, yet do not acknowledge this metaphorical status, and instead operate as though the context were literal. The literalization of this metaphorical connotation implies that a room or person can literally possess a unique energy, like a kind of quality that can be objectively observed and quantified. This is totally nonsensical, and completely incompatible with the true definition of the term. Beyond this, such energies are often referred to as being positive or negative. This is yet another unfortunate linguistic coincidence. In the English language, numbers that are greater than zero are referred to as positive numbers, and numbers that are less than zero are referred to as negative numbers. Because energy can be quantified, and because it is possible for potential energy to have a negative value, depending on how we define certain parameters within a system, we will frequently discuss energy quantitatively using positive and negative numbers. In a totally separate and unrelated linguistic context, positive means good, and negative means bad, as in having a positive or negative experience. For this reason, it is quite easy to take phrases like "positive energy" and "negative energy," remove them from the arithmetic domain in which they have actual meaning, and paint an unrelated connotation upon them, so as to imply that an object can possess "good energy" or "bad energy." This is an unfortunate consequence of allowing words to convey multiple meanings in a language, but as long as we can diagnose the coincidence that acts as the source of the confusion, we can disarm the misuse.

> *TL;DR—The metaphorical usage of energy has been literalized, which muddies the term and misrepresents the field of physics.*

Purveyors of pseudoscience will not stop at the claim that some object, like a molecule, can have "negative energy" with baseless implications. This will always be accompanied by other popular buzzwords. Some of these other words also have a legitimate place in science, such as vibration. Atoms vibrate. Chemical bonds vibrate. This means that they wiggle, and they do so with a particular frequency, which simply means a certain number of wiggles per second. But again, it is often implied that there are "good frequencies" and "bad frequencies," when the word frequency, just like energy, only means anything in a strictly quantitative context. Beyond this, related pseudoscientific claims are made that have no corresponding legitimacy in science. This would involve the evocation of an aura, or spirit. Any claims involving an alleged spiritual world must be placed squarely in the realm of the supernatural, and the supernatural is, by definition, unscientific. Science expressly concerns itself with the natural world, and there is no evidence that anything exists beyond the natural world, or that some aspects of the natural world are somehow unknowable by science. Therefore, anyone who spreads such claims can be rightfully dismissed at minimum as unscientific, and at most fraudulent. In general, because the term is regarded as vague and malleable, energy is popularly evoked when someone has no basis for what they are saying, as an attempt to loosely tether a claim to what would sound like established scientific principles.

A common example would be those who claim a scientific basis for their belief in the afterlife, citing conservation of energy as their reasoning. They reason that our energy cannot be extinguished, and thus must live on in another form when we die. But what energy is this? Upon death, all methods of energy production expire, as there is no longer a two-way exchange of material with the environment. All metabolic activity ceases, and we simply become food for other organisms. The atoms that comprise our bodies are indeed energy, but those don't disappear. They continue to exist, and go on to form other things, whether living or non-living, which could arguably offer some spiritual satisfaction. But people who make this argument are not referring to energy in the form of atomic matter. They are referring to "energy" while implying a spirit, or other immaterial concept, and there is absolutely no evidence that such a thing exists. Does pointing out this disconnect specifically prove that there is no afterlife? No, it doesn't. It simply demonstrates that there is no scientific basis for believing in one and using scientific terminology incorrectly does not lend credence to such speculations.

We will build upon this later when we debunk an array of pseudoscientific fads, but before we get there, let's continue exploring energy as a concept. With what we now know, what can we say about the sources of energy that mankind has utilized over history? Well, the most accessible source of energy available to us is our own bodies. We can move objects. Of course, this fundamentally begins at the sun, and the solar radiation that reaches Earth. We call this light, which is another type of energy, and plants use light to produce chemical energy in the form of carbohydrates through photosynthesis, as we have discussed. Then animals such as ourselves eat plants, so we receive the chemical energy stored in carbohydrates through metabolism. This energy is freed in our bodies by metabolic pathways, and is used to generate ATP, which is required for muscle contraction, allowing us to exert mechanical energy in the way of lifting or pushing an object to change its kinetic or potential energy. But manual labor is quite tedious, and tough on the body, so we eventually had to come up with other solutions, beyond just making animals do our work for us. We quickly became clever enough to take advantage of wind and water in the way of sails and waterwheels, to push our boats and grind our grain. But after centuries of this way of life, the epic transition known as the industrial revolution came about, which required the perfection of the steam engine. This utilized coal as a fuel source, which was burned in a furnace. Burning things wasn't a new practice, we had long burned wood to draw heat from the fire. But this was a new application. When burning coal in a steam engine, chemical energy is released through a combustion reaction, which is used to heat up water to its boiling point, thus producing steam. Steam is a gas, and the pressure produced by this gas is able to pump a piston back and forth. Components that are connected to the piston can then be used to do mechanical work. These steam engines were used to power trains and factories all over the world, making the industrial revolution possible.

The steam engine represents a paradigm shift in energy production.

As times progressed, combustion remained our favorite strategy for vehicular power. We found better sources of chemical energy than coal, namely petroleum. This can be refined to produce gasoline, which in one form or another, fueled internal combustion engines in our cars, planes, boats, and even rocket ships. But we also continued using wind and water, inventing things like hydroelectric plants, which use water from a river to generate enough electrical energy to provide electricity for a whole state. Power lines went up and put everyone on a grid. But the progress of power didn't stop there.

Hydroelectric plants are an example of harnessing the awesome power of nature.

Next came the concept of nuclear power. This is categorically different from the other energy sources we have mentioned thus far. Any form of combustion is based on chemical energy, or the rearrangement of atoms in some fuel to produce gaseous products, along with a release of energy by virtue of the rearrangement. But nuclear power is different. Nuclear power utilizes matter itself as the energy source, converting some of the mass in atomic nuclei directly into energy, as outlined by Einstein and his most famous equation.

The first type of nuclear power that was utilized involved nuclear fission. This is where a very large nucleus splits into smaller nuclei, typically using uranium as a starting material. Uranium is very large as far as atoms go, with well over two hundred protons and neutrons in the nucleus, and many nuclei of this size are tremendously unstable. The strong nuclear force drops off much faster than the electromagnetic force, even over the tiny diameter of an atomic nucleus, so the strong nuclear force holding the protons and neutrons together has difficulty competing with the electromagnetic repulsion between the positively charged protons. A nucleus above a certain size will therefore have the tendency to spontaneously break apart, which we call nuclear decay. This is exactly what happens in a fission reactor, when we bombard uranium-235 nuclei with neutrons. One nucleus splits apart, producing two smaller nuclei, and several more neutrons, which then collide with other uranium nuclei, and so forth, in a chain reaction.

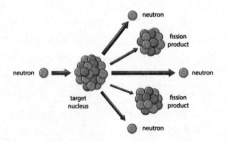

Nuclear fission involves the fragmentation of very large nuclei.

Every time a uranium nucleus splits apart, a tiny amount of the matter in the nucleus, about a tenth of a percent, is converted into energy, which given enough uranium, ends up being so much energy that it can fuel a power plant with much greater efficiency than any form of combustion. How much more, precisely? Nuclear fission releases well over a million times more energy per gram of fuel than the combustion of oil or coal. This energy is then used in a familiar manner, to boil water and power a steam turbine, which in turn produces electricity, like a souped-up version of all the older models. And it is only a minuscule fraction of the total mass that is converted to energy, which gives you a vague idea of how powerful $E = mc^2$ really is. It is so powerful that it has been weaponized in the past, such as with the nuclear bombs utilized in World War II. This is one reason nuclear energy has such a bad reputation. The other reason is that the products of the fission reaction involving uranium-235 are unstable and highly radioactive, so they must be carefully contained.

Unfortunately, there were a number of disasters at nuclear power plants in the twentieth century. We have probably all heard of the Chernobyl incident, involving a nuclear power plant in Ukraine whose reactor exploded in 1986. This was an immense catastrophe in which significant amounts of radioactive material escaped containment, causing several local casualties, and a huge number of delayed deaths from cancer all over Europe. Fission is indeed dangerous business, and detractors of nuclear power are justified in citing these disasters as the basis for their distrust. However, nuclear power also includes fusion, which is another extremely promising technique that is dramatically different from fission. Distrust of fission often spills over onto fusion, and unjustly at that. To understand why, we must first get a better understanding of precisely what radioactivity is.

In short, a radioactive substance is one that emits high-energy particles, and it is the case that these particles can do damage to biological tissue. To understand why this happens, let's return to our very first lesson regarding atomic structure. As we recall, an atom is made of protons, neutrons, and electrons, and the protons and neutrons are in the nucleus. While atoms of a particular element have a fixed number of protons, they can have differing numbers of neutrons. Atoms of a given element with differing numbers of neutrons are called different **isotopes** of that element. So for example, all carbon atoms have six protons, and when naturally occurring they can have either six, seven, or eight neutrons. This leaves us with three natural isotopes of carbon, which are carbon-12, carbon-13, and carbon-14, with masses that are derived by adding the numbers of protons and neutrons for a particular isotope.

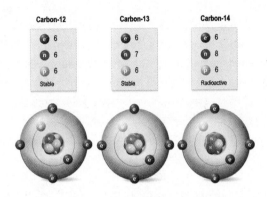

The three naturally occurring isotopes of carbon differ
in their number of neutrons.

Almost all the elements have more than one isotope, and
different isotopes have differing degrees of stability. Carbon-12
is extremely stable because it has the same number of protons
and neutrons, which is desirable for small atoms. Carbon-14,
on the other hand, is not quite as stable, because it has too
many neutrons. Once in a while, a carbon-14 atom will alleviate
this instability by undergoing a type of nuclear decay, which
in this case is called beta decay. One of the neutrons will emit
an electron, as well as another particle called an antineutrino,
which we won't discuss for the sake of simplicity. In doing so,
the neutron becomes a proton. Since one of the eight neutrons
became a proton, of which there were six originally, there are
now seven protons and seven neutrons in the nucleus. This
makes it an atom of nitrogen-14, which is the most stable
isotope of nitrogen, given the equal number of protons and

neutrons. The enhanced stability of the product is what drives this nuclear reaction, which is made possible by another force, called the weak nuclear force. The discrepancy in stability is also what dictates the frequency with which this decay will happen, which we measure in terms of half-life. The half-life of carbon-14 is around 5,730 years, which means that given some sample of carbon-14, it would take that many years for precisely half of it to decay into nitrogen-14. This actually has tremendous application, because we can measure the ratio of carbon-14 to carbon-12 in certain objects to figure out how old they are, given that the ratio changes over time in a mathematically reliable way, and that carbon-14 is constantly replenished in the atmosphere due to cosmic rays.

There are several different types of nuclear decay, because there are different reasons why a nucleus can be unstable. One reason is that it is just too big, as we mentioned earlier. The positively charged protons will eventually blast the nucleus apart from the inside, offering a rough physical limit to how big nuclei can be. In fact, uranium is the heaviest naturally occurring element. All of the elements with an atomic number greater than that of uranium are strictly man-made, formed in particle accelerators by smashing smaller nuclei together at unimaginable speeds. The heaviest element ever synthesized at the time of the writing of this book is oganesson, with an atomic number of 118. This enormous nucleus has a half-life of less than a millisecond, so it exists for but a moment, and then it's gone. We may be able to apply even more power and make even larger nuclei in the future, but the laws of physics will dictate that they will be increasingly unstable, with shorter and shorter half-lives. So it is the case that all of the naturally occurring elements in the entire universe are known and catalogued on the periodic table, with atomic numbers of every integer from one to ninety-two.

When large nuclei decay, they typically do so by alpha decay, which is where a little chunk of the nucleus called an alpha particle, made of two protons and two neutrons, is emitted, with the rest fragmenting into smaller pieces. But for biological systems, which don't contain these elements with very large nuclei, the main reason why a nucleus would be unstable is the unfavorable proton to neutron ratio we mentioned, and depending on whether there are too many neutrons or not enough neutrons, a different type of decay will result to generate a more stable nucleus. Every single type of atomic nucleus has its respective half-life, whether this is a millionth of a trillionth of a second, or a million trillion years, and the half-life is what determines how frequently a substance emits these high-energy particles during any decay process.

It is the emission of these high-energy particles that makes radioactive substances so dangerous. If one of these particles collides with one of the bases in DNA, it can chemically alter the base in a way that results in mutation. More generally, they can simply tear biomolecules apart by stripping electrons away from the atoms they collide with. So a highly radioactive substance with a very short half-life, that is continuously emitting enormous amounts of these high-energy particles, can be absolutely deadly. But the key thing to understand is that essentially everything is radioactive. It is not just uranium or the other nasty stuff in fission reactors. There are radioactive isotopes for every single element in your body, with varying degrees of abundance, which means that there are atoms of carbon, oxygen, and nitrogen inside your body that are decaying right now. These are doing harm to your cells and your DNA. The difference is in the half-life. The most radioactive isotope in your body is potassium-40. This is far less abundant than the more stable isotope, potassium-39, but it is present in tiny amounts, and about five thousand of these nuclei decay per second within the average person. A highly radioactive substance decays millions of times faster, which produces more damage than a biological organism can handle. So just as we learned that every substance is toxic to varying degrees, the same can be said for radioactivity. We are bombarded by the high-energy byproducts of nuclear decay on a daily basis, from any and all objects. It is only those highly radioactive substances that overwhelm a biological system with more radiation than it can withstand.

We can now graduate from a pseudoscientific conception of radioactivity. This phenomenon is not a man-made property held exclusively by nasty synthetic chemicals. It is not the essence of a magic elixir that bestows turtles with the capacity for speech and ninjutsu, as beloved as such tales may be. Touching a radioactive object doesn't make you radioactive, like the vampire's kiss. In truth, everything is radioactive to some degree. It is simply the result of protons and neutrons all over the universe bound in every numerical combination imaginable, to produce the various isotopes of all the elements, and the less stable of those combinations taking action to become more stable.

> **TL;DR—Radioactivity is not a property that can be transferred from one object to another. It relates exclusively to the instability of certain atomic nuclei.**

With this understood, we are now ready to appreciate nuclear fusion. While this still qualifies as nuclear energy, as we are manipulating atomic nuclei, it is the polar opposite of fission in terms of the strategy that is employed. While fission involves splitting huge nuclei apart, fusion involves fusing tiny nuclei together. There are two critical things to understand about this process. The first is that it produces a greater energy output than fission. Every time protons and neutrons fuse together, a small fraction of their mass is converted into pure energy. This is precisely the same process that powers any star, such as our sun. Within the sun, hydrogen nuclei are whizzing around with such great speed that when they collide they can fuse, in a three-step process, to form helium, and the energy this fusion produces is what generates the outward pressure that keeps the star from collapsing under its own gravity. This immense energy output radiates outwards in all directions, perpetuating the incredible temperatures required for fusion to occur, as well as emanating through space to reach earth, making life possible for the past several billion years. The year of this publication, 2020, marks the centennial of the discovery of this process, made by Arthur Eddington in 1920. Prior to this achievement, we had no idea what made the sun shine.

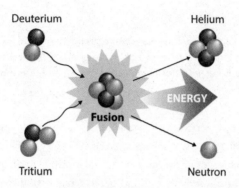

Nuclear fusion involves the synthesis of small nuclei like helium from even smaller ones, like deuterium and tritium, which are isotopes of hydrogen with one and two neutrons, respectively.

The second thing we must understand is that fusion does not carry any of the risks that are characteristic of fission. There are no large, unstable nuclei involved. There is no fissionable material or radioactive byproducts. We are taking isotopes of hydrogen, the lightest element there is, and smashing them together to make helium, just like what happens in the sun. This takes an incredible amount of energy to achieve, given the sun-like temperatures that must be attained, and there are associated challenges regarding fuel containment. Beyond conceptual challenges, early efforts to generate a fusion reaction ran at a considerable energy deficit, producing far less energy than was put in, and acting strictly as a proof of concept. But remarkable advancements have been made, pursuing approaches that utilize lasers, or powerful magnetic fields, and a structure utilizing the latter approach called a tokamak is now getting closer and closer to a net positive energy output, meaning more energy out than in. We are mere decades away from being able to construct nuclear fusion plants that could single-handedly generate more than enough power than our current civilization would ever need, and the public needs to be aware of the tremendous potential for this technology.

> *TL;DR—Fission and fusion are completely different processes, despite both falling under the category of nuclear energy.*

This is what most people imagine when they think of nuclear energy.

Because fusion is a type of nuclear power, it is difficult to present it to the public without conjuring up images of cooling towers, biohazard symbols, and hazmat suits. But it can't be stressed enough that those are associated with fission. Fusion has essentially nothing in common with fission. In fusion, we manipulate atomic nuclei to produce tiny stable nuclei. In contrast, fission involves splitting up huge unstable nuclei. Unlike fission, in fusion there is no dangerous waste, no risk of runaway reactions, no risk of nuclear fallout, and therefore little reason to ignore the promise of this technology. Widespread implementation of fusion power plants would render fossil fuels obsolete, and in the long term, figuring out how to put such a reactor on a spacecraft could be our ticket to exploring distant stars. Fusion is the power of the sun in your pocket. Many would say it's too much power for our species to wield, but there is no basis for this sentiment other than fear. Furthermore, this attitude acts as a disavowal of heritage. We came from stars. Every atom in your body, other than hydrogen, was fused in a star. In the immortal words of Carl Sagan, we are "star stuff." What could be more poetic than a collection of atoms brought into existence by nuclear fusion, assembling over billions of years into a pattern of sentience so wise, that it learns to master the process by which its own material substance came to be?

Fusion is not the only card up our sleeve. The twenty-first century is just getting underway. Before it is through, we will see a complete global transfer to renewable energy sources, which essentially means those that don't run out in the short term, like oil inevitably will. This includes fusion, since the hydrogen that is used as fuel is the most abundant element in the universe. But this also includes solar power, wind power, genetically engineered biofuels, and others that have yet to be introduced. There are brilliant minds all over the world that are devoting their lives to making these technologies a reality and advancing the human race in the process.

There is no inherent limit to human potential. All limitations are self-imposed, by those who are unwilling to learn, those who are unwilling to act, or those who deliberately limit progress because it is not in their own self-interest. The greatest tragedy imaginable would be for humanity to fall short of actualizing its potential. To have the innate drive and ingenuity required to explore the universe and commune with it, but to fall short of that destiny because of the pettier aspects of our nature. We have no way of knowing precisely what challenges are to come, but knowledge is all the ammunition we have, so let's forge ahead and gather a bit more.

CHAPTER II

# An Equation for This and an Equation for That

Mathematics is not a particularly popular subject. Upon being presented with mathematical symbols and equations, many people simply shut down and think about something else until the nuisance goes away. But that's a shame, because there are deep truths in math that can enrich one's appreciation of nature. Mathematics is the language of the universe. It is the great unifier. If we were ever to meet an alien species, their methods of communication would be unintelligible to us, but the mathematical principles they utilize would be just like ours. That's because mathematics is the one realm of inquiry that can provide definitive, inarguable answers to questions about reality. In fact, in a nutshell, math is specifically the study of questions that have definite answers, unlike the subjective areas of inquiry we often concern ourselves with. When we say that the angles in a triangle add up to 180 degrees, we know we are right, with a supreme certainty that can never be applied to anything that is exclusively of the natural world. But fortunately, there are mathematical relationships that are inherent in the universe, which help us decode its ways. Its constituents and properties abide by the rules of math, and they did so before humans came up with any symbols to describe these rules, or before humans even existed in the first place.

All of this is to say that mathematics is part of the fabric of reality. It is not an approximation of reality; it represents the deepest layer of reality. It is our subjective personal experience that is the approximation. The way we experience the universe is limited in arbitrary ways. Why can we only directly perceive a tiny sliver of the electromagnetic spectrum? Why can we only see visible light, and not X-rays or radio waves? Why can some animals directly sense earth's magnetic field, and others can't? Why can some animals use sonar, while others can't? Humans can't do these things, so what else are we missing out on? The list is quite long. We can only conceive of three spatial dimensions. We can't see molecules or anything smaller. We can't logically fathom the immensity of space. Why is it that our senses are so severely limited in directly perceiving the universe and its fundamental nature? It's because they didn't evolve for that task. The ability to see X-rays was not a priority as mammals were evolving, because the majority of the light that hits earth is in the band of frequencies that we call visible light. It's not inherently easier to perceive than other wavelengths, we just evolved the capacity to perceive it because that's what was there, and the ability to perceive one's spatial environment is a useful adaptation, as is dictated by natural selection.

> **TL;DR—What we directly perceive is only an approximation of reality.**

There are so many aspects of the universe that are confusing to us. But mathematics is a tool we can use to transcend our limitations. For example, quantum mechanics tells us that tiny things are both particles and waves at the same time. We can't imagine what something that is both a particle and a wave would look like, if that even means anything. But we can write equations that describe it and predict its behavior to a remarkable degree. General relativity tells us that the three spatial dimensions we are familiar with are warped around a fourth wherever mass exists. We can't depict this visually, or even conceive of it adequately in our minds, as it is beyond our perception. But again, we can write equations and do mathematics until we make some sense of it. It is our perceptions that are the approximation. It is our perceptions that fall short in describing the universe. Mathematics is what describes the universe on the most fundamental level. When we come up with math that does this effectively, we simply have to do our best to interpret the math as best we can and be satisfied with our limitations in that regard.

We didn't always apply mathematics in scientific inquiry. Math and science used to be completely separate constructs. To the ancient Greeks, math was a language that described an imaginary realm of physical perfection. In nature, there were no perfect circles or spheres, flawless right angles, or any other geometrical constructs. This was only possible in the minds of the gods. Science was not empirical at that time. Very little effort was made to collect actual data. To give them the benefit of the doubt, it's not immediately apparent that nature is mathematical. To look at birds flying, rocks falling, or fire burning, and discern that mathematical equations can describe the whole show, it takes quite a lot of insight. That's why this acknowledgement did not fully come to pass until the time of Galileo Galilei, an Italian scientist and inventor from the sixteenth century. He was the first person to demonstrate that terrestrial motion is determined by mathematical equations, by using some simple but elegant experimental apparatus. He had some qualms with the prevailing views regarding terrestrial motion, derived primarily from Aristotle many centuries prior. Among other things, this included the notion that heavier things fall faster than lighter things. Galileo wanted to know precisely how fast things fall, and how that changes as they fall.

He set about constructing some clever experiments. This included all manners of ramps. He reasoned that it is difficult to discern things about bodies in free fall, since they move too fast. But rolling a ball down a ramp would produce similar motion, as the ball is still essentially falling down, just at a much slower rate. He constructed the ramp in such a way that a ball would ring a series of bells on its way down. To measure the time elapsed between each ding, he didn't have anything like a digital clock, but he was able to use things like pendulums and water clocks to figure out the precise acceleration of the ball, and this value could be repeated over and over again. With these tools, he determined the constant acceleration of falling objects. He also overturned Aristotle's assertion regarding velocity's dependence on mass. Though it may be a myth, Galileo is reported to have dropped balls of differing masses off the Leaning Tower of Pisa, to demonstrate that they fall at precisely the same rate regardless of mass, and therefore hit the ground at precisely the same time.

Of course, all of these facts can be demonstrated with incredible precision using modern technology, but it is astonishing that Galileo was able to do it with ramps and balls. This contribution was so significant, that it marks the birth of modern science. Prior to Galileo, science was relatively indistinguishable from philosophy. The recognition that natural phenomena are mathematical and quantifiable sparked a scientific revolution.

> *TL;DR—Modern science began when science was integrated with mathematics.*

While Galileo was working out terrestrial motion, another man named Johannes Kepler was deciphering the secrets of celestial motion. Although Copernicus had proposed heliocentrism a century before, revealing that the sun is the center of the solar system, geocentrism still largely prevailed, which was the notion that everything in the universe revolves around the earth. However, Kepler used the newest, most precise data regarding the positions of celestial objects over long periods of time to refine the Copernican model, showing it to be consistent beyond reasonable doubt. The geocentric model had become a mess. In order to match observations, astronomers had to utilize the concept of "epicycles," whereby planets sat on little orbits that were themselves on an orbit around the earth. The model had to be constantly refined, and there were gaps in its explanatory power. Kepler not only confirmed heliocentrism, but he demonstrated certain aspects of planetary orbits in three laws. He discovered that in going around the sun, planetary orbits are actually elliptical, rather than circular. And he derived equations that offer very precise mathematical predictions regarding the areas swept out by a planet during its orbit, and the periods of time associated with this activity. Kepler's model could be used to predict the future locations of objects with much greater precision than existing models, and this type of approach quickly became the essence of modern science. Models that make precise, quantitative predictions, which are then repeatedly verified, are regarded as scientifically sound, and are therefore assumed to represent some fundamental aspect of reality.

With Galileo having quantified terrestrial motion, and Kepler having quantified celestial motion, it was not long before someone was able to unify these realms, and that person was

Isaac Newton. Newton's list of accomplishments is extensive, but perhaps his most impressive work was in formalizing the concept of gravity. The myth about getting hit in the head by an apple in the garden probably isn't true, but wherever the inspiration came from, he had the stunning realization that the force which makes things fall to the ground and the force which makes the planets go around the sun are one and the same. He was able to derive a law of universal gravitation, whereby the attractive gravitational force exerted between two massive objects is equal to a gravitational constant, G, times the masses of each object, divided by the square of the distance between their centers. This applied to falling bodies and orbiting bodies equally, and in one stroke of genius, Newton had united the divine celestial sphere with this earthly realm, demonstrating that they operate under the same rules.

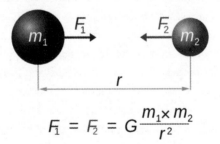

$$F_1 = F_2 = G \frac{m_1 \times m_2}{r^2}$$

Newton's law of universal gravitation.

The philosophical ramifications of this were immense. It meant that now, more than ever before, the universe can be known and understood. We were no longer just stargazers, watching and wondering. A thread of inquiry began that would eventually produce a space-faring civilization. Newton's law of universal gravitation is the foundation upon which we can calculate the trajectories of probes and send them through the solar system, knowing precisely how and when they will reach their destinations.

Newton also developed three laws of motion, and it is the case that these laws, combined with the one on gravitation, are so fundamental that all of Galileo's and Kepler's work on motion can actually be derived from them. So what were these laws of motion? The first one outlines the concept of inertia, which is a resistance toward being accelerated, a property held by any massive object, and in proportion to its mass. This is often summarized by stating that an object at rest remains at rest, and an object in motion remains in motion, unless acted upon by some force. This may not seem revolutionary, but consider the second part of the statement. We all know that an object at rest doesn't just spontaneously start moving. But objects in motion do come to rest on Earth. When you roll a ball, or throw a rock, or induce any motion on any object, we do see it come to a stop at some point. This makes Newton's first law unintuitive. He was able to see that motion would perpetuate if there was no friction. When a ball rolls on the ground, its kinetic energy gradually dissipates because of friction with the ground and the air. Every minute interaction costs the ball some of its momentum, until it eventually comes to a rest. But motion in outer space is not like that. Space is nearly frictionless, since there is no ground, no air, and essentially no other stuff.

Furthermore, there is no nearby source of gravity to influence an object's trajectory. So while a rock that is thrown up in the air on earth will fall back down and come to a stop because of the gravitational force constantly acting upon it, a rock that is thrown in deep space will not. Whatever velocity is imparted onto that rock, by virtue of the force applied, will be retained indefinitely. This is the way motion works in the universe, it is motion here on Earth that is the special case, because of the additional variables provided by local forces, the atmosphere, and other components.

The second law can be summarized by an equation, $F = ma$. Here, F is force, m is mass, and a is acceleration. This applies quite broadly to any force imaginable, from pushing something with your finger, to gravity, to any other force or apparent force. This also means that for any given force, if you increase the mass the force is exerted upon, the acceleration exhibited will be smaller. And if you decrease the mass, the acceleration will be greater. This is why it's easier to push something that is light than something that is heavy. And to recapitulate a previous point, this also specifies that any acceleration requires a force to produce it. Finally, looking at applied forces, the third law simply states that for any force that is applied, there is a reactionary force that is equal in magnitude and opposite in direction. This is also somewhat counterintuitive, but necessary for understanding the motion of objects. It explains why when striking a nail with a hammer, the hammer stops at the point of impact with the nail, because the nail exerts a reactionary force back against the hammer. However, the nail is still driven into the wood, because the only force acting upon the nail itself is the force from the hammer.

Looking at all of Newton's laws together, we can understand how gravitation provides the same acceleration to all falling bodies. Although a more massive object is attracted to the earth with greater force, it also has greater inertia, or resistance toward being accelerated. This fact can be mathematically derived by recognizing that the F in the law of universal gravitation is the same as the F in $F = ma$, so we can set the expressions equal to each other, and cancel out the mass of the object. This mathematically proves that the acceleration imparted upon a falling object in Earth's gravitational field does not depend on the mass of that object, only on the mass of the Earth, and the distance of that object from Earth's center of mass. The culmination of this body of work represents an understanding of the universe that is still used today. We use this Newtonian paradigm of physics to make predictions about cosmic events. When we predict exactly when and where a solar eclipse will occur, down to the second, and within a square mile, nothing is required beyond Newtonian mechanics. Our mastery of the mathematical nature of the universe is revealed every time such a prediction is demonstrated to be accurate, and to an unthinkable degree of precision. This is the power wielded by a system of science-based on mathematics.

To go from Aristotle to Newton is to witness a transformation in thought. There is no doubt that Aristotle was one of the most brilliant minds of the ancient world, but without having adequately applied empiricism, he made many errors. He relied heavily on rationalism, or what we might call "common sense" to draw conclusions about the nature of the universe, applying only pure thought and not experimentation. As we now know, this is a terrible idea. Common sense has no access to the molecular world, because we did not evolve with the ability to directly perceive molecules. So when Aristotle proposed that everything consists of four elements, those being earth, wind, water, and fire, this seemed quite logical to him and his contemporaries. We now regard this framework as archaic and ridiculous, because we are familiar with the periodic table of the elements. But how would someone in Aristotle's time be able to ascertain that the atmosphere contains diatomic nitrogen molecules, diatomic oxygen molecules, and a bunch of other substances? There was simply no way. It took the work of many brilliant people over the latter half of the second millennium, each building upon the work of the previous generation, to slowly build up actual, empirical knowledge. Modern knowledge is based on inquiry, experiment, and data analysis. Common sense rarely aligns with the results of scientific inquiry, and anyone who has any experience with science knows better than to cast their personal musings as actual science, let alone above actual science. This is a major lesson that the public needs to learn, or we will never transcend our current status as a terribly suggestible and easily influenced populace.

*TL;DR—Common sense has no place in science.*

To really drive this point home, let's get a little more backstory on Kepler. Initially, when he was refining the heliocentric model, but before he stumbled upon his laws, he had a preconceived notion of what the orbital radii of the planets should be. He was deeply religious and felt that the solar system should be arranged in a perfect way, fitting of the divine being that assembled it. He proposed that the planets orbit the sun in perfectly circular orbits, and that the distances at which the planets orbit are fixed in ratios that outline the dimensions of what are called Platonic solids. These are three-dimensional shapes constructed with regular polyhedra as faces, which include the tetrahedron, cube, octahedron, dodecahedron, icosahedron, and so forth. The specific geometry of these shapes isn't terribly important. What is important is to understand that Kepler was attempting to project his own will, his bias, in this case a religious bias, onto the physical world. He believed that the number seven was divine, and therefore that the seven planets that were known at the time represented divinity. He believed it would be inconceivable that the planets would be orbiting the sun in any way other than a perfect mathematical ratio. He insisted that the distribution of the planets reveal the mind of God.

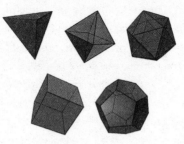

The Platonic solids.

But as any good scientist must do, Kepler allowed his worldview to bend in order to match the data. It was undeniably clear that the planets follow elliptical orbits, not circular ones. And as much as it must have pained him to admit, the planets are not distributed throughout the solar system in some geometrically privileged way. They are strewn about at random, as we now know is the case in every other planetary system we can observe. Kepler had to let go of his desires regarding the way the universe ought to be, as this is outweighed by the way the universe truly is, when the mathematics tells us quite clearly as such. We should learn a lesson from Kepler in this regard, because we all have a penchant for whimsy, to one degree or another. It is far too easy to trick ourselves into adopting a narrative because it feels true, or because it caters to some fear, concern, or desire that we have in life. Advertisers know this, and they exploit our suggestibility. Alternative industries play our minds like a fiddle, identifying the stories and archetypes we are predisposed toward believing, reinforcing them, and manipulating us into adopting a narrative that suits their agenda. These narratives may feel like they must be true, just like Kepler's Platonic solids did to him. But they are not empirical, they are not scientific, and they simply are not reflective of reality. We have already spent some time making this point, but it remains an important notion to reiterate in the context of mankind's graduation as a whole, from ancient superstition to data-driven empiricism. It is one that we should all mirror in our own personal development. The universe does not care what we wish of it, it simply is as it is, from the tiniest particle to the largest galaxy, and everything in between. The mature individual will admit this fundamental truth and make peace with it as early in life as possible.

*TL;DR—Reality does not bend to meet our expectations.*

Before moving on from this historical analysis, there are a few more key advancements to outline. The contributions put forth by figures like Galileo, Kepler, Newton, and other contemporaries represented a revolution. It was a paradigm shift in human thought, potentially the greatest in the history of mankind. Once we recognized that reality could be understood through the language of mathematics, there was no turning back, and we began to successfully explain all the assorted natural phenomena in existence. The laws of thermodynamics helped us understand energy, heat, and spontaneity. Gregor Mendel outlined the basic principles regarding heredity. James Clerk Maxwell elucidated the nature of electricity, magnetism, and light, which became known as electromagnetic radiation, hence unifying these formerly disparate concepts in four simple equations. And then, just as it seemed that the field of physics had become all-encompassing, the next revolution of titanic proportions announced itself. It was the quantum revolution. This word, "quantum," will be the focus of a large portion of the next chapter, so let's define it now.

Physical reality, at the end of the nineteenth century, appeared deterministic. That is to say, we had equations that governed the motion and behavior of all objects and systems. We could perform calculations regarding these objects and systems that reliably predicted their behavior. This clockwork universe of billiard balls moving and colliding in precise, mathematically reliable ways led us to a conception of reality that is exemplified by "Laplace's demon," which embodies the articulation of determinism voiced by French mathematician Pierre-Simon Laplace. In so many words, he mused that because the laws of motion are so precisely understood, if some entity were to know the precise location and momentum of every particle in the universe at any instant in time, they would then be able to derive from that instant the location and behavior of all of those particles at any time in the past or future, in essence becoming omniscient. Since biological organisms are but particles, this view of rigid causality would bring any inquisitive mind to question the concepts of free will and morality, so the philosophical implications run deep. But this is also precisely the notion that was shattered as a result of the quantum revolution.

Quantum theory, from the perspective of the public, is shrouded in mystery. It is typically interpreted and spoken about in an almost mystical tone, as though it were akin to the occult. This is for a number of reasons. First, it is damn near impossible to understand. True comprehension is reserved for highly trained physicists and mathematicians, who can solve exceedingly complex equations. As for the rest of us, we can only strive to understand it on a conceptual level. The problem is that the explanations themselves are mathematical in nature. Verbal explanations tend to be fairly hollow, and therefore easily misinterpreted. However, for reasons that are rather unclear from a sociological standpoint, the public tends to be principally curious about the frontier of scientific fields, in particular physics and astrophysics. The public wants to know about dark matter and quantum tunneling, rather than the work of Newton and Galileo. This is unfortunate, because these frontier areas are the ones that we have the lowest chance of comprehending in an even remotely sophisticated way without mountains of specialized knowledge, let alone basic knowledge. Nevertheless, our desire to commune with the boundary between the known and the unknown drives us to interface with these concepts anyway, and when we do this, we bring lots of baggage to the table. We bring our own fears and desires, and our own biases about how the universe ought to be, just like Kepler did. And the worst part of all is that there is no shortage of media outlets that know exactly what we want to find in the magic mirror of the quantum, and they dish it out to us like birdfeed.

In the next chapter, the final chapter of this book, we will concern ourselves with debunking pseudoscience of all varieties, and a large portion of it will be devoted to so-called "quantum mysticism." In a nutshell, this is the general equating of quantum theory with magic, fantasy, and spirituality, in a way that is completely unsupported by empirical data, and therefore unjustified both rationally and philosophically. This is done to sell products and treatments that fall under the alt-health umbrella, which is the primary target of this book, and this is where we will see all of our efforts come to fruition. A basic understanding of both molecules and energy will enable us to see through the woo-woo, so that we can reject one of the biggest fads in pseudoscientific circles today. But before we get there, we have to develop some foundational knowledge regarding what quantum theory actually is. I know I just mentioned that this is nearly impossible without complex math, but we are going to try anyway. We've come this far, what is there to lose?

To put things as briefly as possible, quantum theory is about two things. The first of these is quantization, or an explanation of reality in terms of **quanta**, which are teeny tiny indivisible things. The second is about probability, and framing reality in a probabilistic manner, in stark contrast with the deterministic universe that was understood at the turn of the twentieth century. Let's deal with the quanta first, as they lend their name to the theory.

There are four fundamental forces in the universe. Those are the strong nuclear force, weak nuclear force, electromagnetism, and gravity. Forces had always been, and still can be in certain contexts, discussed by way of fields. A gravitational field, an electromagnetic field, and so forth. What quantum theory seeks to do is replace these fields with interactions among elementary particles, or quanta. In this way, the strong nuclear force is mediated by the exchange of particles called gluons. The weak nuclear force is mediated by W and Z bosons. The electromagnetic force is mediated by virtual photons. And gravity is allegedly mediated by gravitons. Only the last of these remains completely hypothetical, all of the rest are based on very sound science, and many of these particles have been confirmed to exist in particle accelerator experiments. The question is, why do we need these quanta?

This all began when we noticed some gaps in the classical theories. There were observations that could not be explained by existing models, and we found that one such observation could be solved if we assumed that energy was quantized. This meant that energy cannot exist in absolutely any amount imaginable. Energy cannot be delivered from one place to another with any value from a continuous spectrum of values. Instead, there exists some smallest amount of energy possible, and then multiples thereof. It's kind of like hard currency. In America, the smallest coin is the penny. You can't spend less than one penny. You can spend many pennies at once, if something costs more than one penny, as indeed everything does. But you can't spend less than one penny, and you also can't spend fractions of a penny. Anything you purchase must cost some multiple of one penny. When we assumed the same thing for energy, we found that we were suddenly able to explain a lot of observations, even though it was not immediately clear what this meant regarding the nature of the universe. But we forged ahead with the idea, eventually applying it to space and time, which we now consider being constructed of fundamental, indivisible units as well. Einstein first made it onto the map by extending this idea to light, introducing the concept of the photon. A photon is a particle of light, whereby light had been exclusively thought of as being comprised of waves until that point. Light exhibited particle-like behavior in certain situations, where waves would not suffice, and this was the beginning of **wave-particle duality**. This is the notion that light is both a particle and a wave at the same time, a concept that was subsequently extended to encompass all matter as well, most notably the electron. We began to describe electrons not as mere particles, but also as waves. And in particular, we began to regard them as waves of probability density. This is where the probabilistic nature of the universe became apparent.

Werner Heisenberg then put forth his famous uncertainty principle. As we mentioned, particles had always been regarded essentially as little spheres that would behave sort of like billiard balls, though much tinier. We viewed them in the same deterministic way as we viewed balls on a pool table, assigning them a discrete location and a precise momentum at all times. But then we realized that particles such as electrons are not like billiard balls. They have wave-like character. Therefore, they do not possess a well-defined location and momentum simultaneously. The uncertainty principle describes this, relating the product of the uncertainty in a particle's position and the uncertainty in its momentum as relating to a constant, called Planck's constant. What this meant was that the more we know about the location of an electron, the less we know about what it's doing. And the more we know about what it's doing, the less we know about where it is. We can only state the probability with which an electron will possess certain values for these parameters. Erwin Schrödinger developed quantum mechanics, including his eponymous equation, which allowed us to calculate details regarding the atomic orbitals that electrons must reside in within an atom, which revolutionized both physics and chemistry. And after all of that, there was no turning back.

> *TL;DR—Quantum theory tells us that nature is fundamentally probabilistic.*

That's a lot to take in, so let's pause for a moment and clarify a few things. First, when we talk about being unable to discern both the position and momentum of an electron at the same time, we have to understand that this is not a technical limitation. It is not because we don't have the technology to determine them simultaneously, it is because the electron literally does not possess these parameters in a discrete manner simultaneously. If it did, it could be regarded strictly as a particle. But it is not just a particle, it is also a wave. We are able to force such a quantum system to collapse and provide us useful information about one parameter or another, but never both, because that simply is not in its nature. It is also not the case that a quantum system somehow reacts to consciousness, or knows that it is being measured, or any of the other misinterpretations that are so commonly uttered. The plain fact of the matter is that electrons, photons, and all other quantum systems are both particles and waves, and it doesn't matter that it doesn't make sense to us. It doesn't matter that we can't visualize it, or rationalize it, or compartmentalize it. It's the best representation of reality that we have, and thus seems to simply be the way it is. And there is plenty of evidence that this is indeed the way it is, as any chemist can attest to, since various theories regarding so-called hybrid atomic orbitals and molecular orbitals, which are mathematical functions related to probability, are absolutely required to explain a wide variety of chemical phenomena. This is how these systems behave, period. When empiricism and common sense are at odds, empiricism must always be allowed to emerge the victor, no matter what the psychological toll.

Next, we must try to comprehend what this means for reality on the most fundamental level. The universe is probabilistic in nature. We cannot define with certainty every parameter of a quantum system. We can only specify the probability with which it will exist in a certain state. Each state has its own respective probability, calculated by complex mathematics, and if we zoom out and look at a macroscopic system, the states of all the particles in the system will exhibit a distribution that flawlessly matches the relative probabilities, precisely as dictated by a subfield of physics called statistical mechanics. In this way, we can say that determinism, which describes the billiard ball world we are familiar with, is an **emergent property** of the probabilistic quantum world. Everything we see is made of tiny quanta, which are probabilistic in nature, each obeying the rules of quantum mechanics. But when you have a system that is made up of enough of these quanta, such that reliable predictions can be made regarding the system as a whole, Newtonian mechanics emerges, forcing bulk matter to behave in the familiar way.

These are the main lessons that quantum theory has to offer. All the forces are mediated by tiny particles called quanta, everything is both a particle and a wave, and nature is fundamentally probabilistic on the smallest scale. Don't worry if these things don't immediately make complete sense to you. It would be shocking if they did. The key thing to absorb is that this is what quantum theory says, and it does not say a whole lot else. It does not say that consciousness influences matter. It does not abuse unrelated scientific terminology. It does not even offer concrete interpretations of its own mathematics, of which there are several. There is so much that the public erroneously infers from this theory that we simply had to go over the basics regarding what it actually says, to set the stage for the debunking to come.

Before we move forward, I would be remiss if I skipped over an enormous misconception held by the public, one which fits best here, as we wax philosophical. This misconception regards the definition of the word theory. In the common vernacular, a theory is a guess. It implies uncertainty, a lack of finality, a work in progress likely to be discarded. This is not accurate, and such a definition can be more legitimately attributed to the word hypothesis. A **theory** is a model that is used to correlate data and make predictions. A theory takes an event from over to the left, a measurement from over to the right, an object from up here, and some data from down there, and explains all of them elegantly with a single model or set of equations.

This theory must be rigidly falsifiable by way of the concrete predictions that it makes. This means that the theory must say: "For this theory to be correct, if I do X, then Y should happen." Variations of such predictions are tested many times. If they turn out to be incorrect, the theory has been falsified. If the predictions are always correct, the theory is corroborated to some degree, and if increasingly specific predictions continue to be correct over a long period of time, to high degrees of quantitative precision, we can say it has been corroborated beyond reasonable doubt. Its utility becomes apparent, and it becomes an indispensable part of the scientific body of knowledge. If an idea does not make such falsifiable claims, such that its validity cannot be rigorously tested, it does not qualify as a theory, and it is not science.

Another misconception is that if a theory is found to be true,
it becomes a law. These are totally separate constructs, and
theories never become laws. Laws are mere summaries of
observation. They are statements regarding what happens,
with no explanatory power whatsoever. For example, Newton's
law of universal gravitation is an equation that quantifies the
gravitational force between two objects, and it is extremely
useful. But it does not explain what gravity is. Only a theory
can do that, like Einstein's theory of general relativity, which
describes gravity as a ramification of the curvature of spacetime
around massive objects. We don't necessarily have to
understand what that means, we just have to understand that
theories do not imply uncertainty, and theories never become
laws, nor are they below laws on the pecking order. If anything,
they sit at the top, due to their explanatory ability. And there
they remain until a new theory comes along that can explain the
data more completely, should such a theory ever arise, never to
demolish the old one and take its place, but simply to relegate
the old to the realm where it remains successful, and explain
phenomena in a grander and more pervading way.

Take for example, atomic theory. There was a time where we
postulated the existence of atoms. This was then formalized into
a theory, which immaculately explained certain observations
regarding the ways that different substances react with one
another in specific numerical ratios. Because of this, the
theory was thoroughly corroborated, and of course at this
point we know for a fact that atoms exist, because we can do
chemistry. But it is still called atomic theory. There is no "law of

atoms." The existence of atoms has simply been corroborated unimaginably far beyond reasonable doubt, such that nothing in science makes sense without them. The same can be said for the germ theory of disease. In a time when disease was still thought by many to be caused by demons or bad odors, some proposed instead that certain tiny living organisms were responsible. This was controversial at the time, but we studied bacteria in microscopes, began to learn about pathogens, developed the field of microbiology, and now it is common knowledge that bacteria and viruses cause certain diseases. But it is still called the germ theory of disease. No amount of corroboration alters the definition of a theory.

So we can now see the difference between a theory, a hypothesis, and a law. But unfortunately, the collective consciousness has obfuscated these terms unrecognizably. The ramifications of this are dire, and I don't think that's an overstatement. A thousand times a day, someone utters the phrase "that's just a theory." This is a denouncement, a declaration of alleged uncertainty, which almost always results in the rejection of firm, well-corroborated science. The Big Bang cosmological model is a theory that is consistent beyond reasonable doubt, though it is rejected by many purely on the basis of their misunderstanding of the term. Evolution by natural selection is also criticized in this manner, typically without comprehending that evolution is an undeniable process which is observed on a daily basis. Natural selection is the theory that seeks to explain how evolution propagates, and it does so immaculately. Biologists do not doubt in the slightest that all biological life evolved from a common ancestor around four billion years ago. The supposed uncertainty of theories is used as a scapegoat to project one's own baggage onto science and reject it with no basis. This has large-scale social ramifications. If we are unable to curb the narrative that science is hopelessly uncertain, it paves the way for the narrative that science is out to get us, or at least that the road to scientific progress is rife with peril. In actuality, ignorance and complacency are the true dangers. This attitude ensures that the next pandemic, or ecological disaster, or asteroid will wipe us out in an instant. Scientific knowledge is our safeguard against what nature throws at us. So we must never utter "that's just a theory" in common parlance. If we have to use a scientific term in this manner, it should be hypothesis. The denigration of the theory happened slowly due to misuse, just the way that "literally" is now taken by some to mean "figuratively" for the same reason. The only recourse is to stop using the word in this colloquial context altogether, such that the false connotation, as embedded as it is in the collective consciousness, will eventually fade away. This is one small thing we can all do to make a difference.

> **TL;DR—Theories are not inherently uncertain, and they never become laws.**

Theories are powerful. They allow us to look at seemingly disparate natural phenomena and realize their intimate connection. They allow us to make concrete predictions regarding the origin of the universe, the origin of life, and the nature of reality. When these predictions are verified to high degrees of precision, our knowledge grows, and by extension, our command of technology. Indeed, the emergence of science as a rigidly mathematical and quantifiable practice is the single most important turning point in our quest to understand everything around us, as it demarcates the philosophical musings of the ancients from the repeatable, empirical, technology-producing science of today. An understanding of what this kind of science looks like makes it very easy to sniff out any pseudoscience that doesn't make the cut, and we are now truly ready to do some sniffing, so let's go ahead and see what we turn up.

CHAPTER 12

# To Debunk Is Divine

If you've made it this far, it's time for your reward. We are going to take our newfound knowledge in chemistry, biochemistry, biology, and physics, and we are going to use it to tear through a variety of fads, misconceptions, and hoaxes that are rampant in our culture today. Some of these phenomena are harmful to society. They endanger lives and cause artificial rifts in the populace that others can use to gain power. Others are mere flights of fancy that subsets of the population engage in with very little broader harm as a result, apart from being a drain on their bank accounts. In debunking these narratives, the aggressiveness should be proportional to the harm that is being done by the narrative. Exposing pseudoscience should only be done in a bellicose manner if that pseudoscience specifically harms others. This is because, apart from being a bit cruel, attempting to strip someone of their beliefs without provocation will often entrench them further in their delusions. Those who did not use logic to arrive at a position will not respond to logic-based arguments for the abandonment of that position. But when we are directly confronted with pseudoscience, no matter how benign, it is the duty of the educated individual to shine a light on ignorance. Even a passing interaction with a peer, or a stranger, where pseudoscience is presented, is an opportunity to strategically influence that person's thinking, or at least plant a seed of doubt. The odds of success are low, and the strategy utilized may differ depending on the personality and skill set of the individual, but if we intend to transform our society into one that is wise and capable of critical thought, we must challenge ignorance wherever it is encountered, as every drop counts.

With each pseudoscientific narrative we disarm, and with each source of misinformation we discredit, we get closer and closer to reaching our potential as a society. Whether this means exposing a snake oil salesman or conveying some scientific facts to a friend in a nurturing manner, I challenge the notion that such activity is futile. I reject the idea that people can't be convinced of anything. I scoff at the proposition that everyone should just stay in their lane and keep their heads down. Inaction is a cop-out. We are here to build a world together, and the collective consciousness responsible for its sculpting is a chaotic system of people and thoughts. We all play a role, and we all have some amount of influence, so let's start using that influence wisely. It is possible to change the way people think, one must simply be skilled and persistent. Let the debunking begin.

To avoid pulling a muscle, let's warm up a bit. Pseudoscientific claims are made left and right regarding water. First up is the phenomenon called "hexagonal water." This is described by its peddlers as water molecules that arrange themselves in a hexagonal pattern, with eight molecules in a ring formation. This structure is said to have unique properties with "exciting health ramifications." What they do not mention is that this pattern of water molecules goes by another name. Ice. Water freezes in this pattern, among others depending on the precise conditions, because that's how it can maximize hydrogen bonding. The only thing this could accurately be describing is ice cubes, which sadly do not have health ramifications beyond a mild local anesthetic. This pattern absolutely is not retained when water is a liquid, because that defies what a liquid is. In the liquid phase, molecules move past each other constantly, like balls in a ball pit. So apart from fleeting clusters that last for less than a trillionth of a second, hexagonal liquid water is not a thing.

Similar claims are made regarding energized water, alkaline water, oxygenated water, and so many other kinds of super-special water, all of which are desperately trying to convince you that they are better than normal, distilled, pure water. But much to their chagrin, energized water doesn't mean anything. Alkaline water implies a basic pH, which by definition is not pure water. This would be instantly neutralized in the acidic environment of the stomach and is not advantageous in any way. And oxygenated water is just water with bubbles in it, which is not useful, since we are not fish, and oxygen enters our bodies through the lungs, not the digestive tract. We also can't "change the molecular structure" of water by whispering words of love or hate into it. Beyond the sheer absurdity of the claim, if you were to change the molecular structure of something, it would not be that molecule any longer. Images of water crystallizing in different arrangements, called polymorphs, are the result of manipulating ambient conditions to affect the rate of freezing. The images are then arbitrarily paired up with claims regarding exposure to certain genres of music or particular verbal messages. It's a hoax. All these bizarre claims really do is dismiss the relevance of practically unlimited clean water from a tap, a very recent development which is one of the most incredible achievements in the history of humanity, one that the underdeveloped world can only dream of.

> *TL;DR—All "special" waters that boast health benefits are hoaxes.*

Now let's aim for a more significant target. An enormous collection of pseudoscientific concepts falls under the umbrella of "quantum mysticism." The ethos, or rather the angle taken by this fad, which first picked up steam in the 1980s, can be summarized as follows. Quantum physics is confusing, so it's magic, therefore if magic can be framed using the terminology of quantum physics, that magic is real. Pushers of this narrative tend to be people who sell books or services. The key to this narrative is its philosophical allure. Religion has historically given people a sense of purpose, interconnectedness, and divinity. As the civilized world grew largely secular, this necessarily meant withdrawing from fundamentalist religious beliefs. There are those who have the tendency to trade one kind of religiosity for another, and quantum mysticism offers such a belief system, containing all of the transcendence and empowerment without any of the burdens of organized religion. Personal desires can be manifested, not through prayer but through meditation. The instinctual yearning for permanence is satiated by tying deep truths to the practices of the ancients. And according to them, all of this is backed up by science, because in the twentieth century, science figured out that reality is magic.

In actuality, quantum mysticism is what you get when you combine two highly contradictory traits. On one hand, a self-image of being science-minded despite supreme ignorance toward science, and on the other hand, a proclivity toward fantasy. It is the epitome of pseudoscience, because it makes absurd claims and presents them as being conclusions that are logically derived from accepted physics. But the interesting thing is that it simultaneously stands in defiance of Big Science, and in particular, Big Medicine. It is costumed as primeval truths from a time when humans were aligned with planetary energy, despite their thirty-year life expectancies, that have recently and suddenly been obscured to make way for Big Business and their profits. In short, it's just more of the same childish narrative we have been dealing with this whole time.

*TL;DR—Be skeptical of anyone using the word "quantum" who is not a physicist. They don't know what it means, and they assume that you don't either.*

Let's get a bit more concrete with some of the language that is put forth. I live in Los Angeles, California. The enclave of Santa Monica appears to me to be the epicenter of quantum mysticism, harboring holistic healers and psychics of more flavors than the ice cream section at the grocery store. For example, examining the website of one local organization that shall remain nameless, the owner is described as a reiki master teacher, author, sacred sound alchemist, quantum healer, and meditation instructor. I see tabs on the home page entitled Chakra Shop, Reiki, and Quantum Resonance. Clicking on Chakra Shop brings me to a page with crystals for sale. These are called Pure Consciousness Crystal, Multi-Dimensional Light Body Crystal, Magic Wand of Universal Power, Sacred Octagonal Shield, and other such blatant frivolities. I will add that they are not cheap. And what do they do, you ask? Well, the Pure Consciousness Crystal purportedly can be called upon to assist you with all forms of self-healing, especially with tapping into your Highest Self at all times. It is said to bring expansion, upliftment, and activation to your energy body to heal the present and keep you connected to your Radiant Source, continually raising your consciousness and vibrations. They help to assist in personal and global ascension. They have their own consciousness and are always accompanied by the Crystal Angels and Masters. They connect you with Higher Dimensions of the light matrix by activating your Merkaba, which is your personal light body. They can be used to energize water, food, and beverages, and create a protective aura around your body, rooms, and buildings. Wow! All that for only $122? I'll take ten!

Healing crystals are just pretty rocks.

My friends, I genuinely could not make this stuff up if I tried. One could have a stroke in attempting to count the number of baseless assertions in this description, which I even edited down for brevity. Let's first list the things it references that do not exist. An "energy body." What's that? Is it just the body, because all matter is energy? Then calling it an "energy body" is like saying bread pizza, or animal dog, or internet computer website. If that were the implication, it would be redundant albeit legitimate. But it certainly is not the implication being made. It implies the existence of some pseudo-energy that it refuses to define or validate. Radiant source is meaningless. Global ascension is meaningless. Light matrix is meaningless. Merkaba, or "personal light body" is meaningless. Now for the baseless assertions.

**The crystal allows for self-healing.**

How? By what mechanism?

**It can raise your consciousness.**

What does that mean? How do you quantify consciousness?
How do you get more of it? Does the crystal manufacture
neuronal cells and implant them in your brain?

**They have their own consciousness.**

Really? Can you demonstrate that? Does it respond to stimuli?
Does it answer questions? Does it say "ouch" when you poke it?

**They are accompanied by Crystal Angels and Masters.**

What are those? Where are they? Can I see them? Can I touch
them? Why can't I see or touch them?

**They can be used to energize food and beverages.**

What does that mean? What kind of energy? Kinetic energy?
Does the crystal heat up food? Can I cook my dinner with it?

**It creates a protective aura.**

What's that? Can I see it? Can I measure it? Can I detect it? If
there is no way I can interact with it so as to detect it, how does
it interact with other things to protect me? How does it work?
How do I know it's not some nonsense you made up out of thin
air to sell me pretty rocks at a 5,000 percent markup?

You get the picture. It is said that extraordinary claims require extraordinary evidence. This website is a veritable smorgasbord of extraordinary claims, and yet precisely zero evidence is offered anywhere. Because there is no attempt whatsoever to justify any of these outlandish claims, it is not even remotely scientific, or reasonable, and the only justification someone would have in believing it is because they want to believe it.

> *TL;DR—Crystals don't have magic powers.*

So where does the science come in, such that we could even label it pseudoscience, as opposed to pure, unadulterated fantasy? There are a few buzzwords that any quantum mystic will cling to. These are energy, frequency, and vibration. We talked about energy for an entire chapter, so we are all set there. The usage of frequency and vibration is easily elucidated.

All electromagnetic radiation has a particular frequency. This includes radio waves. For those still listening to radio stations in the car, each station has its own characteristic frequency. This allows for the car to tune in to a particular station. Each station has its own genre of music, and most people have genres that they prefer. If you find a station you like, this is a preferable frequency for you to tune into, and this has been taken by metaphor to imply that there are "good frequencies" and "bad frequencies." Of course this is ridiculous. There is nothing inherently good or bad about any frequency, we just

arbitrarily choose certain frequencies to transmit information
through space that can be decoded by a receiver to play music
of rock, country, rap, or any other genre. Your body doesn't
have a frequency. You can't get to a better or worse frequency.
Anyone who uses this word in this context has no idea what
they are talking about. The same can be said for vibration.
As we learned, atoms and molecules vibrate with particular
frequencies. This refers exclusively to how quickly they are
vibrating. Despite what The Beach Boys would tell you, there
are no good or bad vibrations in a literal sense. It's just another
instance in which the metaphorical usage of a term has been
literalized and bastardized.

In an attempt to entrench themselves deeper within accepted
physics, they will attribute their misuse of these terms to a
particular quote by Nikola Tesla. This is generally referenced
as follows:

*"If you wish to understand the universe, think in terms of
energy, frequency, and vibration."*

As you can see, we have unearthed the genesis of this holy
trinity of buzzwords. When Tesla said this, he was talking about
waves. Waves carry energy. Waves have a frequency. Ambient
heat energy causes molecules to vibrate. This talk of waves
applies to electromagnetic waves, or light, and mechanical
waves, like sound, and even elementary particles, which
are also waves. Our ability to perceive the universe around
us depends on these waves and how they interact with our
bodies. Therefore, our perception of reality is determined by

waves. That's all. He was not referring to some elusive and unquantifiable vibration of the mind or soul, nor the existence of good or bad energy, nor was he implying any of the magic that people commonly evoke when referencing this quote. He was a builder of machines, not a mystic. Reducing all of reality to the word "vibration" is childish and ignores well-understood physical and chemical principles. The intent in referencing this quote is abundantly clear. It is an attempt to ground magic in the words of an esteemed physicist, and therefore physics itself, by appealing to people who do not understand what these words mean.

This is the self-image that quantum mystics sell to consumers.

But the half-hearted attempt to root itself in science is not the essence of this movement. Like almost all other pseudoscience, it's about the narrative. It's about offering a version of reality that people want to be true. When people visit a website that depicts a fit, lovely woman in an impressively contorted meditative pose, with a colorful array of glowing orbs superimposed up her spine, and galaxies swirling in the background, this is the image that is being sold. It's the same as the rugged outdoorsman and the equally rugged beer he is drinking. It's the same as the elegant debutante in a ballgown and her enchanting fragrance. It's marketing. People want their anxieties to congeal into serenity which gives way to perfect control over their lives and minds. People want to feel relevant in the universe and connected to everything in it. People want superpowers which allow them to influence events through their thoughts. But even the most skilled meditator is not a god. We can't fly around the universe with pure will, we are stuck down here on Earth. And we don't have superpowers, despite what the mystics say. This movement is pseudoscience for those who wish to feign compatibility with actual science. It is modern day wizardry, and nothing more.

> **TL;DR—Mystical connection to the universe is a marketing tactic.**

Let's put this a bit more concretely. On that same website, we mentioned a tab for reiki. What is this? The website describes it as a "vital, high-frequency healing energy." It is a "powerful energy healing modality, much like acupuncture without the needles." Well, acupuncture is the practice of sticking a bunch of needles in a person, so without the needles, you are left with nothing, and that's what reiki is. Reiki is nothing. Reiki involves hand motions over the patient, sometimes touching and sometimes not, which is supposed to channel "healing energy" from some "universal source." So essentially, it involves someone waving their hands along someone's body and playing pretend. It is faith healing, just with the Christian god swapped out for Asian mysticism. Rather than evoking a deity, reiki is dressed up in symbology and terminology from the Far East, which is meant to act as an anchor of legitimacy. It presents itself as an ancient tradition, but it isn't. It's not more than a century old, a fact that is regularly obfuscated. This website goes so far as to say that "it simply requires those receiving the treatment to be open and accepting of the loving, gentle energy." The translation is that if you don't believe it will work, it won't work, because indeed it doesn't work, but rather operates by placebo effect, which doesn't work if you don't believe it will work.

This is what practitioners and clients of reiki imagine is happening.

If there was any legitimacy to this technique, it could be demonstrated. Why not take 100 participants, blindfold them, set up the ambience equivalently, but for half of the participants, the practitioner runs their hands and accessories over the body, and for the other half, they keep their hands at their sides. Will the half that didn't actually receive the treatment complain that they don't feel any effects? Oh, by the way, this website also offers remote reiki. So don't worry, the nothing that is happening between their hands and your body can do the same amount of nothing from across a computer screen, as though this shouldn't raise any doubts. Reiki for pets is listed as a service too, just in case you have a dog that also embraces pseudoscience. Welcome to Santa Monica.

This is what really gets me the most about quantum mysticism. Its practitioners spin a narrative about how lost modern science and Western medicine are, how absurd it is that they would ignore these deep truths about reality and the powers within a human spirit. How arrogant scientists and doctors must be to reject such powerful healing techniques, thinking themselves bigger than the forces of the universe! In actuality, this is the precise opposite of the truth. The reiki practitioner is the arrogant one, to the point of delusion. This is a person who believes, given the benefit of the doubt, completely genuinely, that they possess magic powers. They believe that they have access to a tier of reality that others are unable to sense, that they are simply so special, so wise, so enlightened, that they have command over forces that elude the rest of the species. What could be a more narcissistic delusion than this? With reiki, it's a collective delusion, because there is also a client. The client wants to believe that magic exists and pays the practitioner to put on a performance that supports the notion of magic existing. This is the essence of the service. It's money exchanged for the confirmation of a delusional worldview. Somebody goes in, there's a little song and dance, they get a placebo effect, and they leave. So what's the big deal? Is this a crime? Certainly not. And in fact, if you were looking to pay the price of a massage but not get one, then there is no big deal. Reiki doesn't seem to typically be pushed as an alternative to medical treatment. It's a bit more recreational, a drain on your wallet and not your health. But we aren't finished yet.

*TL;DR—Reiki offers nothing more than a placebo.*

Finally, let's tackle this so-called Quantum Resonance treatment that is also listed on the website. Apart from transparently mashing together two scientific terms for no reason, it also combines a delightful little cornucopia of techniques involving being bathed with colored light, placing crystals and other such accoutrements at specific positions on the skin, listening to music, being subjected to magnetic pulses, and other such practices. What are some of the claims that are made regarding the effects of this glorious all-in-one treatment? What is it precisely that this is said to do? Well here is a list of effects that are stated explicitly on the website.

**Creates positive DNA expression.**

What is that? How can these treatments influence
gene expression?

**Relaxes through vibration therapy.**

What is that? What is producing vibrations, and how? Why can't
we feel the vibrations? What effects do these vibrations have,
and how can they be measured?

**Helps autoimmune disorders.**

That's a pretty bold and concrete statement. Have you done a
study? If so, where were the results published?

**Works with cellular intelligence.**

Is that like little molecules with black suits and sunglasses,
staging military coups in nearby Latin American cells?

**Rebuilds the subtle bodies.**

What are those? Organisms? Cells? Molecules? Where are they?
What do they do, and what is it that makes them so subtle?

**Reduction of inflammatory markers.**

Another bold claim. Where's the data?

**Noninvasive and safe.**

Fine. You know what else is noninvasive and safe? Doing absolutely nothing.

**Suitable for all ages.**

Yes, light and rocks are indeed suitable for all ages.

**Homeostasis of the inner and outer field of the body.**

What's the inner field? What's the outer field? What are they comprised of that requires homeostasis? Details, please.

**Better sleep through chakra balancing.**

What is a chakra? Where are they located? What do they do? Can we see them? How do you balance them? What effect does that have on the body?

**Balances pH levels.**

This by definition requires acid-base chemistry. Light is not molecules. So where are the acids and bases, and how are they getting into the body?

**Based on Arcturian and Pleidian star technology.**

I can't think of what this could possibly imply other than that this technology was delivered by aliens from outer space. That's right, aliens crossed the galaxy to bring us colored light bulbs and pretty rocks.

It really is astonishing that anyone could read any of the above and take it seriously. It almost sounds like a parody of itself. But there are people who not only believe that these claims are valid, but that it is the truest form of medicine. There are people who believe that these types of treatments are the secrets to health and wellness, and that Big Science doesn't want you to know about them. This is precisely where quantum mysticism crosses over from absurdity into danger. If an adult client is manipulated into foregoing legitimate medical treatment in favor of light and rocks, it is unethical and immoral. If parents decide to "heal" their children in this manner, again circumventing medicine in the process, it should be treated as a crime. If the child dies, nothing less than negligent homicide has been perpetrated, and the parents should be prosecuted accordingly. The justification offered on the website for these treatments is so blatantly unscientific that any layperson could debunk it for themselves with five minutes of googling if they were so inclined. But people who are attracted to this narrative are not skeptics and have no interest in challenging their presuppositions. When their gullibility endangers others, it becomes a broader problem, and there is no system in place to shield people from these types of lies. These industries are not regulated, so people are not scrutinized or held accountable for their claims the way that other industries are.

> *TL;DR—Mystical healing is dangerous when it supersedes actual medical attention.*

It is important to identify how problematic it is that claims made by quantum mystics require no verification in order to be publicized. If they did, their industry could not survive, because the rare instances in which concrete claims are made, they are indefensible. For example, on the website, next to "Royal Rife Frequencies," it is said that every disease has an "electromagnetic frequency" and that producing an impulse of that same frequency will "reset your DNA and cells" to the proper frequency. These claims are so arrogant in their glaring disregard for scientific principles, that they presume sporadically inserting a vocabulary word from a middle school science curriculum is sufficient in making the claims sound scientific. Simply saying words like DNA, cells, and frequency, as an attempt to ground these practices in science, is ridiculous. Frequency of what? Light? Sound? Vibration? Where is the table listing these frequencies? What is the frequency of Alzheimer's disease? What about jaundice, or arthritis? This is a very specific claim that can mislead someone into bestowing these treatments with medicinal value that they simply do not possess. Imagine if a pharmaceutical company marketed a new drug, stating that it cures Alzheimer's, when it's actually just a capsule full of table sugar. In what world would this be allowed to happen? And yet, this roughly describes how many of these practitioners operate.

As we touched upon when discussing homeopathy, while it is irrefutable that these treatments offer nothing more than a placebo, perhaps it is not prudent to completely discount the value of such an effect. It does have measurable medicinal value in that it can make someone feel better. Beyond this, medical doctors may or may not be interested in connecting with a patient on a personal level, as they tend to be focused exclusively on the physiological basis of disease. Practitioners of alternative treatments probably owe the majority of their success to being affable, and engaging with their clients conversationally, to the degree that the interpersonal interaction has a psychologically therapeutic effect. There is no major transgression associated with a brief chat, a few rituals, and a client that leaves relaxed and cared about. Outrage should be reserved for the instances where specific claims are made that misrepresent the services beyond this highly limited value.

With this analysis of quantum mysticism in the alt-health realm, we referred exclusively to one website, but honestly, it's enough. Other websites advertising similar services offer nothing beyond slightly different word salads. Croutons may be seasoned with a different spice, and the dressing may be a little tangier, but it doesn't matter. It's just the words quantum, energy, frequency, and vibration, adorned with whatever poetic load of garbage that particular perpetrator can conjure. But there is a vehicle behind the transparent motivation of charging money to touch a rock, and it is a lucrative one. Therefore, not surprisingly, quantum mysticism transcends the territory of alt-health and seeps into the world of self-help. Just the way that famous motivational speaker Tony Robbins will tell you about your personal power, overcoming self-doubt, or accessing the right mindset, a prominent mystic like Deepak Chopra will talk about how the universe is a conscious, living being. He will claim that sentience can be attributed to literally everything, from subatomic particles to entire galaxies. He will state that you have control over your genes and the way they are expressed, and therefore can control your own aging process, or other aspects of human biology. He insists that these facts are derived directly from the postulates of quantum physics and are not of his invention whatsoever. The message is clear. Science says that your consciousness creates your reality.

Again, just as with the medical treatments, these are extraordinary, highly testable claims. And yet, no legitimate effort is made to test them. They simply serve as a stage upon which to pontificate. In actuality, the motivation for this rhetoric is clear. The idea of the universe as consciousness is nothing more than an attempt to deify the cosmos. It is the manufacturing of a god for the otherwise godless. Those of us who reject organized religion still possess all of the neural proclivities toward seeking and affirming divinity. From an evolutionary perspective, we want there to be a god. This is what this brand of quantum mysticism offers, the universe as God, in a very literal way, and a connection to this god which allows you to manifest your desires. You have the capacity to create your own reality, and once you decide that you are ready to take this role seriously, buy this book, or attend this seminar, and the universe will open up like an oyster. With all of this mega-manifesting so trivial to achieve that the instructions fit neatly in a paperback, it's surprising that we don't see Chopra winning big on roulette at the Bellagio, don't you agree?

Perhaps it's because lying to impressionable people has been lucrative enough.

So where does this drivel come from? When priests like Chopra claim a scientific basis for their remarks, what is this in reference to? These claims come exclusively from the blatant misinterpretation of key experiments in early twentieth century physics. For example, there is the famous "double-slit experiment." This is where particles are allowed to diffract through two slits and generate an interference pattern on a detector beyond. There are different variations of this experiment, utilizing either light, or electrons, or other particles, but the key factor is that when not interfering with the process, the material appears to pass through both slits, and produce a pattern to be expected of a system with wave-like behavior. Only when attempting to ascertain which slit something passes through, in the case of an individual particle, do we get a concrete answer, thereby affecting the data received at the detector. The pseudoscientific narrative posits that in observing the system, it changed, thus the system reacted to being observed by consciousness. This is objectively false. In attempting to ascertain information about the system, we must physically affect the system. In the submicroscopic realm, observation is not a passive process. You can't just "look" at a particle the way you look out your window. You can't secretly spy on a particle. Detection involves specialized instrumentation and the emission of photons which interact with the system, and thus detection necessarily affects certain properties of that system. It is the physical act of taking a measurement that affects the system, not our consciousness. That's all there is to it. Wave-particle duality is not a mystery, it's math. There is no magic, there is no voodoo, and there is no basis upon which to run around spouting nonsense about how we can create any reality we want with our minds. It's asinine.

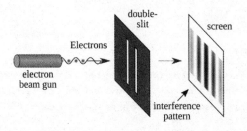

The famous double-slit experiment.

We can truly see how all other leaps of logic made by this paradigm stem from this specific confusion regarding a singular experiment. Anything you hear about the universe being immaterial, or a mental construct, is preposterous. The degree to which different media outlets attempt to disguise themselves as real science varies, with some giving it quite the old college try. These will misrepresent philosophical statements made by scientists, distort scientific data, and present the narrative of the quantum mystic as that of a pioneer, pushing the forefront of physics against resistance from the stubborn backwards thinkers within the field. Just to give it a dash of paranoia, government conspiracies are sometimes referenced. No matter how much intellect goes into the wordsmithing, the narrative of the quantum healer is as hollow as his or her claims, and the motivations are equally transparent. The quantum healer seeks to leverage universal fear regarding disease, mortality, and uncertainty, and offer a magical solution. This magical solution is revealed through highly profitable platforms, like an endless string of books with evocative titles and sunburst covers, reminiscent of the aforementioned yoga babe perched upon her throne of flax seed smoothies.

> **TL;DR—Claims about the universe being conscious, and consciousness creating reality, stem from a deliberate misrepresentation of key experiments in physics.**

So this "quantum manifestation" isn't real. So what? If the narrative is motivational, isn't it the same shtick as Tony Robbins? No. The difference is that Tony isn't saying anything that is objectively wrong, at least to my knowledge. Overly reductive? Sure. A bit vapid? Perhaps. But in the end, if something he says motivates you to make changes within yourself, to modify your own behavior willingly, and you benefit from this transformation, then there's no problem. If you regard the experience as being worth the money you paid, no one can tell you otherwise, because no claim was made about anything other than your attitude, your emotions, and your happiness, all of which is personal and subjective. This is categorically different from something like quantum mysticism, which makes bold assertions about the nature of reality. It takes the guise of cutting-edge physics, when in reality, it exemplifies pseudoscience. Even if this narrative doesn't cause specific harm, it leads society astray. It encourages the masses to entertain magic rather than learning actual scientific principles, presenting itself as the next step in science, such that they can reject everything that came before. In short, it is a major impediment on the pathway to global science literacy.

I'm going to put this as bluntly as possible, because you've been with me this whole way, and I owe you as much. Chakras aren't real. Crystals aren't enchanted. Nobody has magical healing powers. You can't manifest a Lamborghini just by thinking about it. And anyone who says otherwise on any of these points is either infantile or selling you something. The universe does not bend to our will. Life is hard. You can't solve your problems with magic. You can't cure diseases by placing a rock on your skin or bathing in blue light. The people who develop legitimate medicine go to school for a very long time and work very hard to do so, and those who have not gone through this process have no basis with which to challenge their expertise. Just as any child eventually learns, Santa Claus isn't coming to bring you presents, you aren't actually Batman when you play pretend, and Deepak Chopra isn't going to do anything but take your money. Quantum mysticism is one big pile of horse manure, and that's the end of the story.

I wish that this level of scorn could now be put aside, but unfortunately, there is a long list of fads that deserve to be dealt with in precisely the same manner. Fortunately, most of what we just discussed can be transferred over to almost every other form of pseudoscience in one way or another. Take a look at astrology. What does astrology have in common with quantum mysticism? It does not reference quantum mechanics, so the flavor is different, but the recipe is the same. The assertion that modern science is limited and narrow-minded in its inability to recognize deeper truths about the universe? Absolutely. An evocation of the ancients? You'd better believe it. In fact, astrology is one of the only pseudosciences that actually is legitimately ancient. But once again, this fact does not bolster its claims, it merely exposes its lack of substance. Astrology proposes that the positions of celestial objects at the time of one's birth determine the personality of that person, and the positions of these objects continue to influence events on earth every day. But they absolutely do not. Astrology is to astronomy what alchemy is to chemistry. It's a primitive and mystical foundation upon which real science was eventually built.

There are two routes one can take to make swiss cheese of astrology. First, the mechanistic route. Stars are physical objects, and we are physical objects, so if stars influence us, there must be some physical mechanism by which this occurs, which can be detected and studied. Can this be framed by one of the four fundamental forces? The strong and weak nuclear forces operate on the scale of the atomic nucleus, so those are out. Electromagnetism is also out, as bulk matter is electrically neutral. Gravity is the only force that matters on astronomical scales, so can this be the one? Well, if you recall Newton's law of universal gravitation, we can plug in some numbers regarding masses and the extreme distances to the stars, and we find that the gravitational influence of even the closest star besides our own is less than that of the toaster in your kitchen. By many orders of magnitude. So no, astrology does not operate by gravity, or any other fundamental force, not that any of them could conceivably influence your personality in the first place. The rational person would then conclude that there is no mechanism by which this influence could propagate. But the mystic would persist, stating that the mechanism is simply unknown, or even unknowable.

Finding ourselves at an impasse, this leads us to the second route. Astrology makes falsifiable predictions, so do they hold up to scrutiny? Astrology claims that certain people will be a certain way. Are horoscopes accurate? Its assertions would also require that everyone born on the same day in the same region be astonishingly similar. Are they? The answer to both of these questions is a resounding no, as has been demonstrated by countless studies. Professional astrologers can never match horoscopes to their owners in a statistically significant way. People regularly identify with horoscopes that are not theirs, because all horoscopes are vague and flattering. Horoscopes from different sources say dramatically different things. And no matter how vague the horoscopes get, they still only correlate with actual events a tiny percentage of the time. By the same token, studies have been done following hundreds of people born within seconds of each other in the same region over decades. They were born under an identical night sky, so if astrology is to be true, there must be some similarities amongst these people. Instead, there is nothing. Truly, not a single thing.

We can even go so far as to look at sets of fraternal twins. There is no risk of false data here regarding birthplace or time of birth, since fraternal twins are necessarily born nearly simultaneously and in the same room. Are all sets of fraternal twins identical in personality and temperament? Absolutely not. Therefore, astrology is conclusively false. In science, we are interested in what is true. If something is only kind of true, or true once in a while, then it's not true. Newton's law of universal gravitation doesn't work a quarter of the time, it works all the time. With astrology, one can convince themselves of its validity only with a hefty dose of confirmation bias.

> *TL;DR—Astrology is a pseudoscience.*

In the end, astrology is a collection of mystical beliefs that existed prior to empirical science. It came about in a time when we thought the earth was the center of the universe, and that celestial objects were literal gods, traversing the night sky. The entire cosmos revolved around our petty dramas. We had no concept of DNA being the thing that determines what we are, so the stars above seemed like a reasonable conclusion. But constellations are just the result of past generations playing connect the dots with points of light in the sky, creating mnemonic devices to help us remember the coming of the harvest, or the stormy season. It's an outdated worldview. We now know that we are not the center of the universe. We know what planets and stars are. We understand the science that drives them. And the only basis for a persistent correlation between them and us is whimsical delusion. Astrology is like quantum mysticism in that it appeases an immature desire for the whole universe to be intertwined with the trivialities of our everyday lives. We want the entire galaxy to care about who we are and what we do, such that we can feel divine purpose in a purposeless world. In short, it's just another flavor of the same old archetype we have been debunking since page one.

If we were so inclined, we could extrapolate these delusions endlessly. Acupuncture is nothing more than sticking needles in your back. Aromatherapy is just nice-smelling oils. Breatharianism is the insane delusion that one can live without food, or even water. Crystals don't have magic powers. Cupping therapy doesn't do anything but leave a mark on your body. Miracle mineral supplements are tantamount to drinking bleach. Naturopathy is merely unsubstantiated potions for the chemophobic. In reality, almost everything that can be described as alternative, new age, integrated, holistic, naturopathic, or any other buzzword that stands in defiance of "mainstream science," is rubbish. This is the default position that any sensible person must take, and any claim that seeks to elevate a pseudoscientific idea should be greeted with extreme skepticism. Claims can be examined, and, given sufficient quantitative evidence, they can even be validated and adopted into the body of scientific knowledge, but this is exceptionally rare within the landscape of these types of narratives.

> **TL;DR—Skepticism should be the default position
> of any educated person.**

Throughout this book, we have identified a wide array of
fallacious thought regarding scientific principles. If there could
be one thread to connect it all, it would probably be "industry,
bad; science, bad." But another more subtle thread would be
the notion that in opposition toward our inborn inclination to
seek meaning and unity, the sobering impartiality of science
seems to smash that meaning to bits, leaving us cold and
alone. Indeed, this is the type of accusation that a charlatan
would conjure to attempt character assassination upon the
skeptic, provided that they are above simply labeling them a
government shill. But while one may be inclined to resist the
sterility of science because it would seem to eliminate purpose,
this is actually the opposite of the truth. No, the universe
does not revolve around us. The sky is not watching over us
and making sure that everything goes well today. Magic isn't
real, even when you're sick, even when you're lonely. But
what is the conclusion that can be drawn by this decoupling
from delusion? It means that we create our own reality. Not
the way the quantum mystic would have you believe, but in a
concrete manner. Far from being magic, the universe is actually
comprehensible through its physical mechanisms. We can learn
about it, discover its secrets, and use them to improve our
lives. The study of natural phenomena does not detract from
their beauty. On the contrary, this beauty is only enhanced by

the depth of our comprehension, and all the tiers of reality that were invisible before we decided to probe them. And most importantly, there is nothing out there imposing purpose upon us, so we are free to recognize purpose within ourselves, and make of it what we choose. I can't imagine what could be more poetic than this.

EPILOGUE

# Science and Industry in an Educated World

In this book we covered a lot of ground, from the molecular level to the organismal. While we did not go especially deep on any one topic, as one is unable to do when staying as broad as we have, I hope you feel the foundation of a scientific worldview congealing in your mind. Because we now have a rudimentary understanding regarding the behavior of chemical and biological systems, we are ready to read claims regarding phenomena that involve these systems. We are equipped to consider them and determine whether they fit into an already existing framework of self-consistent knowledge. We know how molecules behave, so if we encounter something that contradicts our comprehension, we can immediately reject it. We understand how medicine works in a general sense, so if we come across something that opposes this paradigm, we can resist its manipulative intent. We also have a reasonable sense of how the scientific community operates. With this familiarity comes an immunity to the narrative of science as dogma. We know that it is a dynamic field, constantly restructuring itself, and perpetually transforming the world around us. We know that science simply seeks to understand the universe, and that good science tends to have technological application. This allows us to spot bad science from a mile away.

Of course through all of this, we remain aware of the primary purpose of production in a capitalistic society. Profit is exclusively what drives the gears of industry, but we are no longer so naïve. All of the Big Businesses we are taught to fear certainly fall under this umbrella, but so do all the others, the ones that are so cunning as to present themselves in a different light. The alt-health industry is just as relentlessly profit-driven, and certainly no less duplicitous, than any industry that has ever existed, a fact that is compounded by its complete lack of regulation. Amongst these dueling titans and their dealings, there are us lowly pawns that seek our camp, and who can be led astray in our quest for purpose. Not everyone who defends pseudoscience is a charlatan. Some have simply been fooled into making it part of their identity, and they wave its flag like a birthright. These are the ones we should seek to influence. These are the people who have nothing to gain from the misinformation they propagate and can therefore be converted to a position of reason if we can disempower the narrative they have selected. Take note that this is not a partisan endeavor. Institutions of all political loyalties create propaganda, and every sector of the public is vulnerable to this propaganda in its own unique way. Conservative anti-government and liberal anti-industry sentiments alike, we are being played in great numbers, in the way of artificial rifts that obscure deeper issues.

The challenge of total decoupling from propaganda and misinformation may seem insurmountable, but the stakes are as high as they get. If we can find a way to educate the masses, an unprecedented level of unity will be reached. Imagine a society that decides together how to manipulate commerce through directed voting trends and boycotts. Imagine a society that is able to execute its collective whim to redistribute wealth as it sees fit, rendering predatory economic behavior obsolete. This society would be virtually limitless in its capacity to mold its own future, which would undoubtedly become as close to a utopia as is possible within the limits of physical reality.

How do we get there? We need strategies. I won't pretend to have a true nuts and bolts algorithm to suggest. But I have a sense of some things we can do to help ourselves along. The first strategy is to restructure the virtues we advance as a culture, by way of the media we engage with and the people we elevate to celebrity status. If we truly wish to promote knowledge as virtue, then the people we rally behind should be knowledgeable. They should be wise, intellectual, and compassionate. We should not select celebrities that embody nothing more than beauty, wealth, and materialism, as we currently tend to do. We should build a celebrity class that embodies what we hope to become as a society. So instead of obsessing over wealth, let's embody scholarship, and reward the champions of this pursuit with our sharpest attention. Everyone should be able to name this year's Nobel laureates rather than the cast of the latest Real Housewives of Who Gives a Damn. Instead of materialism and ownership, let's focus on community and allocation. Let's manifest what the 1960s could have been: a true restructuring of the social fabric and a redirection of our deepest passions.

The second strategy is to rehabilitate the image of science and popularize the learning process, while continuing to make education freely accessible and more effective. The more knowledge one possesses, the easier it becomes to accumulate more knowledge, with the end result being expertise. In science, this expertise belongs exclusively to scientists. We have to be clear on the seemingly obvious fact that scientists know science best. Certain outlets would have us believe that the frontier of medicine, astrophysics, and other fields is trivial to understand, with overly reductive content that lures the reader into thinking they have achieved mastery of a concept as complicated as dark matter. But the amount of training and technical knowledge required to comment on the forefront of any field is monumental, and the public must learn to weigh the validity of their position accordingly. When the consensus of the public is not aligned with that of the scientific community, we must make an effort to identify the root of the confusion and help guide the population to a more informed stance. This will be easier if large-scale popularization of learning is successful. It is also the only way that we will be able to collectively reject false narratives and agree on the nature of reality.

What do we stand to lose if we are unable to achieve this? Scientific progress will continue, whether the masses are on board or not. However, if we do not make progress collectively, we will be at the mercy of the elite, and the existing chasm between classes will widen. Before the twenty-first century is through, we will likely see our civilization take its first steps toward manipulating the human genome, curbing the aging process, and colonizing other worlds. We may find ourselves in tricky ethical territory as a species, and the less the public knows about science, the easier it will be for special interests to polarize and divide us. People who are prone toward a conspiratorial mindset are the easiest to control, because they can be fed any narrative that aligns with their predispositions, and they will be made to act against their own best interests. Likewise, we need to have enough of a command over scientific concepts to know when an industry is indeed overstepping its bounds. The key to discerning between these opposing situations is knowledge. We need to know what industries are doing with our science. We need to know how they are doing it, why they are doing it, and exactly how it affects us. If we don't know these things, we become easy targets for misinformation. Industries will capitalize on our ignorance and subjugate us to no end.

We must all do our part. We must hold ourselves to the highest standards when it comes to the content we engage with and promote, placing profundity over inanity. We must undermine and expose those who paint science as hopelessly uncertain, or arrogantly short-sighted, or specifically out to get us. We must resist the infiltration of pseudoscience in the academic space and remind our peers that lies are sold on fears and feelings. We must light a spark and arm the masses with the outlook and the capacity that will enable mankind to transcend its current limitations, and take its place as a wise, technology-wielding, cosmic civilization. Intrepid we have been, and intrepid we must remain, if we wish to make of ourselves what we all hope humanity can be.

# About the Author

Dave Farina has a BA in chemistry from Carleton College and an MA in science education from California State University, Northridge. After years of teaching chemistry in the classroom, he turned his attention to science communication in 2015, focusing primarily on his YouTube channel Professor Dave Explains.

Dave's channel serves as a database of over a thousand tutorials in a wide variety of scientific fields. These educational videos have just enough detail to help struggling high school and undergraduate students while being visually engaging and general enough for any viewer who just wants to learn a few things.

Although Dave is passionate about helping students all over the world reach their academic and career goals, he is even more passionate about bestowing the public with basic science literacy. Anti-science sentiment has been growing in recent years, and it has demonstrated itself to be harmful to society. Dave is devoted to diagnosing and disarming common anti-science narratives and exposing pseudoscience wherever it crops up.

Mango Publishing, established in 2014, publishes an eclectic list of books by diverse authors—both new and established voices—on topics ranging from business, personal growth, women's empowerment, LGBTQ studies, health, and spirituality to history, popular culture, time management, decluttering, lifestyle, mental wellness, aging, and sustainable living. We were recently named 2019 *and* 2020's #1 fastest growing independent publisher by *Publishers Weekly.* Our success is driven by our main goal, which is to publish high quality books that will entertain readers as well as make a positive difference in their lives.

Our readers are our most important resource; we value your input, suggestions, and ideas. We'd love to hear from you—after all, we are publishing books for you!

Please stay in touch with us and follow us at:

**Facebook:** Mango Publishing
**Twitter:** @MangoPublishing
**Instagram:** @MangoPublishing
**LinkedIn:** Mango Publishing
**Pinterest:** Mango Publishing
**Newsletter:** mangopublishinggroup.com/newsletter

Join us on Mango's journey to reinvent publishing, one book at a time.